SO YOU THINK YOU'RE HUMAN?

Felipe Fernández-Armesto's recent books include *Civilizations*, *Food*, and *The Americas*. He regularly presents *Analysis* on Radio 4, and is Professor of Global Environmental History at Queen Mary, University of London.

'One of the most formidable political explicators of our time'
New Statesman

'He makes history a smart art'
The Times

'One of the best historians in the world'
Literary Review

'A superb storyteller'
Independent on Sunday

Praise for *Millennium*

Felipe Fernandez-Armesto has accomplished a herculean task …
I was left wanting to go back and read it all again.
The New York Times Book Review

He makes history a smart art.
Victoria Glendinning, *The Times*

He is in a class of his own for serious scholarship.
John Bayley, *The Spectator*

Stupendously informative and elegantly written.
The Boston Globe

Praise for *Food: A History*

Highly provocative and entertaining … an erudite and
surprising book with many eye-opening pleasures.
The New York Times

A sparkling ramble through history which scatters
countless pointers to new research on the way.
Financial Times

As interesting to the hungry man in the street
as it is to the austere academic.
The Wall Street Journal

Praise for *Civilizations*

A superb storyteller, with a barrel-full of anecdotes and a language as finely textured as many a novelist.
Independent on Sunday

Enthusiastic readers of popular history have come to expect the author of Millennium to deliver a read filled with wonders, important insights, wit and outrageous opinion. A marvelous new work.
Publishers Weekly

An agile writer, possessed of impressively deep knowledge as well as originality. A book full of surprises about humanity's relations with nature.
Booklist

In Memory of
AUGIE

SO YOU THINK YOU'RE HUMAN?

A Brief History of Humankind

Felipe Fernández-Armesto

UNIVERSITY PRESS

Great Clarendon Street, Oxford OX2 6DP

Oxford University Press is a department of the University of Oxford.
It furthers the University's objective of excellence in research, scholarship,
and education by publishing worldwide in

Oxford New York

Auckland Cape Town Dar es Salaam Hong Kong Karachi
Kuala Lumpur Madrid Melbourne Mexico City Nairobi
New Delhi Shanghai Taipei Toronto
With offices in
Argentina Austria Brazil Chile Czech Republic France Greece
Guatemala Hungary Italy Japan South Korea Poland Portugal
Singapore Switzerland Thailand Turkey Ukraine Vietnam

Oxford is a registered trade mark of Oxford University Press
in the UK and in certain other countries

Published in the United States
by Oxford University Press Inc., New York

Reprinted 2009

ISBN 978-0-19-969128-9

Printed in the United Kingdom by
Lightning Source UK Ltd., Milton Keynes

Contents

List of Plates

INTRODUCTION

The Arena of 'Humanhood'

Here is a paradox. Over the last thirty or forty years, we have invested an enormous amount of thought, emotion, treasure, and blood in what we call human values, human rights, the defence of human dignity and of human life. Over the same period, quietly but devastatingly, science and philosophy have combined to undermine our traditional concept of humankind. In consequence, the coherence of our understanding of what it means to be human is now in question. And if the term 'human' is incoherent, what will become of 'human values'? Humanity is in peril: not from the familiar menace of 'mass destruction' and ecological overkill—but from a conceptual threat.

The challenge has come from six main sources. First, primatology has heaped up examples of how like other apes we humans are. It's hard to find any supposedly rational capacity nowadays which primatologists won't tell you is replicated among other apes: language use, tool-making, symbolic imagination, self-awareness—you name it, they'll find non-human apes who do it. Chimps and humans are objectively so alike that an anthropologist from Mars might classify them together—agreeing with human scientists who like to call *Homo sapiens* 'the naked ape' or 'the third chimpanzee', or who call for the boundaries of the genus *Homo* to be re-drawn so as to include non-human apes within it. We are in the most perplexing stage of an argument that has raged for centuries over the differences between humans and other primates ('degenerate men' in many medieval characterizations). Are humans 'naked apes', distinguished from other apes by peculiar physical characteristics, or 'enchanted apes', differentiated by divine inspiration, or endowed by some mystery of evolution with a kind of consciousness other animals cannot attain?

Secondly, the animal rights movement has been remarkably successful in challenging us to identify what, if anything, entitles us to privileged treatment, compared with other animals. Beyond human diversity lies an indistinct frontier between human and animal realms which were once thought to be mutually exclusive. In modern times, the problem of distinguishing humans from animals has inspired some radical solutions. According to Descartes, animals resembled machines, whereas in the human machine there was a ghost. His followers went further: the cry of a beaten dog was no more proof of pain than the sound of an organ when its keys were struck. This was not an inference from observable facts—the plain man would persist in believing, insisted Lord Bolingbroke, in the difference 'between the town bull and the parish clock'—but, as the historian Keith Thomas has said, it could provide a rationalization of the way people treated animals, which would otherwise seem unpardonably cruel. In the present state of knowledge of the continuities between humans and other animals, it is impossible to sustain any account of human nature as sharply differentiated as Descartes'. At the other extreme, one of the consequences of blurring of old convictions about the differences between humans and other animals has been a boost to 'deep ecology'. Peter Singer denounces humans for 'speciesism' and preaches 'equality beyond humanity'. Much objective thinking is on the side of the animal rights lobby. From a perspective informed by science and philosophy, John Gray has asked equally hard questions about what it means to be human in relation to other animals. Meanwhile, all kinds of non-biological criteria for humanhood have been proposed (humans are tool-making animals, animals-with-language, cooking animals, animals-with-self-consciousness, animals-with-imagination, moral animals, and so on) and all, on close scrutiny, prove unsatisfactory.

Thirdly, debate about the moral implications of our self-definition as humans has become inseparable from an issue in palaeoanthropology: how far back in the evolutionary past to distinguish humans from others. In any case, palaeoanthropology has made the traditional limits of the genus *homo* nonsensical. There no longer seems anything definingly special about specimens we class in the same genus as ourselves, and those we relegate to the realms of 'ape-like' australopithecines. Now, as we look back over the fossil record, we see features once thought definingly human—such as bipedalism, big brains, use of tools, omnivorous diet—shared among various species, including some from outside our line of descent. The intensity of the current scholarly battle over the Neanderthals (discussed in Chapter 4 below) reveals the depths of insecurity some humans feel at the discovery that other species can be like us, with similar minds, emotions, and ethical capacities. Arguments over the human status of Neanderthals have been conducted in terms startlingly reminiscent of nineteenth-century controversies about blacks.

Fourthly, in the last half-century or so, biology seems to have changed the balance of the age-old philosophical question about whether species are natural kinds, with essential, universal traits, or merely sets or categories into which we group creatures for convenience. In the present state of knowledge about evolution, it is hard to go on believing that there are any traits which are both general within a species and unique to it. Species have vague, variable boundaries. What makes them species varies from case to case. If species have a unifying feature, we do not know what it is. To belong to one is to belong to a class, not to evince an essence; in a sense, it is a temporary condition, subject to revision, not an eternal and inevitable fate. If we belong to the human species, we do so not because we have any particular qualities, but because that is how we draw the boundaries of our lineage.

Fifthly, artificial intelligence research has stimulated philosophical re-thinking of concepts once key to human self-definition, such as consciousness, reason, imagination, and moral passions. Just as primatological and palaeoanthropological work convinces us that such qualities have evolved and might again evolve in species other than our own, so the quest for artificial intelligence raises the speculation that other beings, of our own creation, might have all the qualities that make us human—and therefore that robots can, in effect, be like King Louie in Disney's version of the *Jungle Book*: king swingers, 'human too'. We could share our world with humans who have human creators instead of human progenitors. It may never happen, but we could be prodded or panicked into re-thinking the nature of humankind before it does.

Finally, genetic research has given us a way of measuring for membership of our species and, simultaneously, of calibrating how much we have in common with all others. Being human has never felt so beastly. And genomics promises or threatens to deliver the project of Dr Moreau: hybrids with human qualities genetically grafted onto non-human stock. How would we classify such creatures? How does the fact that they are conceivable affect our concept of ourselves?

In the twentieth century humankind had at its disposal the most inclusive self-definition ever: in effect, every group was human if members of other groups could successfully breed with it. Yet, despite this inclusiveness, the century was disfigured by the most horrific inhumanities ever recorded. 'New humanism' arose in reaction, while the pace of interculturation outmoded racism and helped to stimulate the drive to make human rights coterminous with the concept of humankind. In the late twentieth century, however, the problem of understanding the concept acquired two new dimensions which enriched its complexities and accentuated its agonies. First, social pressure to license abortion has created a new, effectively sub-human category: the unborn baby (formerly assumed to be fully human and to participate fully in human rights). In consequence, a new quality called 'personhood' has now been invoked to justify the reclassification of the unborn, and the old question of how far humanhood is a biological or a cultural status has acquired a new focus of interest. Secondly, work in artificial intelligence and genetic engineering has made the human future as problematic as the human past. Do we face the 'post-human' future forecast by Francis Fukuyama? Or, given the unresolved history of the concept of humanity, can we simply yank at its elasticity and prolong the debate?

This book attempts to confront these problems by looking back at how our concept of humankind has developed historically. For here is another paradox: those of us who think we are human feel utter confidence in our human identity and our ability to recognize it in others; we hardly pause to congratulate ourselves on the breadth of our views, sensing common humanity in specimens of our species in spite of differences of colour and culture. Yet our present concept is a recent contrivance: most people in most societies for most of history would have been astonished by such an all-encompassing category. Most of them, indeed, would have had difficulty in understanding the word 'human' or finding an equivalent for it in their own languages, except as a way of designating members of their own group. To them, outsiders would belong to some other, alien class, along with beasts or demons. The present limits of our concept of humankind are not obvious and not universal. They have been attained as the product of a long, hard struggle in the Western world to find a way of understanding humanity which embraces communities formerly excluded by racism and ethnocentrism, while insisting on a clear distinction between humans and non-humans. In the present state of debate, and in the light of available knowledge, this looks increasingly like an incomplete and, perhaps, a non-viable quest. This is not the time for a conclusive or exhaustive study of it. What follows is an essay in the history of 'humankind', not of humankind without the inverted commas: the concept, not the people it denominates. It is, of course, too big a subject for the format of this book, but since, despite its importance, it has never been attempted before, and the relevant literature is scattered and fragmentary, it seems worth presenting an outline sketch which may help to inspire further work, rather than attempting an exhaustive enquiry.

The fact that we take our concept of humankind for granted is, to me, a cause of concern: it is a form of complacency which makes us ill-equipped to face challenges. I suspect, moreover, that we are wrong if we think it needs no further stretching. It needs to be scrutinized for deficiencies. Palaeoanthropologists who want to encompass more hominids in the category, primatologists who want to redraw the limits of the genus *Homo* in favour of chimpanzees, moralists who deplore the exclusion of the unborn and the moribund from some human rights are all, in their ways, tugging at the limits of the concept: it may still have surprising elasticity in it. The story of the broadening of the concept of humankind—the story outlined in the pages that follow—is not over yet. The question, 'What does it mean to be human?' (or the follow-up question, 'So who is human?') now evokes among us different responses from those offered in the past or in cultures other than our own, but it remains as difficult to answer as ever.

We have, it seems, never ceased to be apes; yet we aspire to be angels. How far have we really got along the evolutionary road? How far have we got to go, before we have genuinely included the whole human community, and reached a viable frontier between humans and others? Perhaps the quest is doomed to be interminable as every scientific advance blurs formerly convincing distinctions.

CHAPTER 1

THE ANIMAL FRONTIER

Problems of Human Self-definition

Many of our favourite stories are anthropomorphic. The fables of Aesop or La Fontaine are morally convincing because the characters are crows and mice and foxes, though the lessons are intended for men. The satires of Aristophanes would not bite so deeply into the foibles of humans if they were not disguised as birds and frogs. In children's stories, personality traits which would seem shallow and uninteresting in human characters gain charm from animal guises. The muppets would not be half so amusing if they were presented as humans, rather than as fluffy pigs and cuddly bears. Satan is plausible as a serpent; C. S. Lewis re-cast Christ as a lion. Totemic myths tug at our sympathies. Our imaginations blur and traverse the frontier between humans and other animals.

But there is another indistinct frontier, hard to negotiate, between anthropomorphism and zoomorphism. When we put human words in non-human mouths and human emotions in the breasts of beasts, what is being misrepresented? Is it human nature—or the natures of the animals we invoke as vehicles for our stories and scepticism about ourselves? Are we admitting that we are like them or wielding irony in the service of self-differentiation from them? In spite of the terrible condescension with which we admit beasts to our mythologies as honorary humans, we keep to ourselves. The distinction between human and non-human nature is sacred to us—a taboo it is unnatural to challenge. Whoso lieth with a beast, he shall surely be put to death.

We think we are human. But when did we start to think that? When did people begin to put members of other communities in the same category as themselves? When did they begin to draw radical distinctions between beings identifiable as human and those which could be relegated to different—and usually 'lower'—orders of being? The longest-debated frontier of human identity is that between humans and other animals. This chapter is about why that frontier is contested; why it is impossible to fix; and why, instead of defining it better, our growing knowledge of it only makes it seem more indistinct than ever.

Subjects of the Animal Kingdom: The Impossibility of Self-exile

We look in the mirror and recoil from the beast we see there. It is tempting to suppose that what differentiates our species from others is our obsessive urge to classify ourselves apart from the rest of creation. Do dolphins put themselves in a category distinct from that of other creatures? Do viruses think of themselves as lords of creation, or lions as 'kings of beasts'? We shall never know for sure: in spite of the enormous scientific apparatus at our disposal, there is a level at which some other species disturb us by seeming to understand us far better than we understand them—when they fly from us in fear or cajole us with coquetry. On the face of it, it seems that the distinctions we draw are not apparent to some other species, who are capable of adopting members from other communities with surprising ease: most birds willingly host intruded cuckoos; wolves suckle wolf-children. Domestic pets, it is true, do not lose their sense of being dogs or cats or chimps or whatever: they still sniff out their own kind for purposes of breeding or establishing territoriality, but at the same time they seem to classify themselves—as it were—as honorary humans and to behave as integrated members of their host communities. They eat human food, take part in human recreations and economic activities, guard human infants as if they were their own, respond to human speech with sometimes surprising degrees of understanding, and, often, as far as physical constraints permit, imitate human behaviour. Invited to sort piles of photographs, apes reared by humans commonly put their own pictures into the 'human' pile, while putting those of broadly similar creatures, such as monkeys, into the non-human pile.

Throughout recorded history, almost every supposedly distinguishing feature by which humans have identified themselves and differentiated themselves from other creatures, classified as non-human, turns out to be mistaken or misleading. Let us leave aside, for the present, all the criteria formerly invoked to exclude groups now considered as human—criteria, for instance, of colour, body shape or size, cranial formation, nature and distribution of body hair, and so on—for these are subjects of other chapters in this book, and review only the indicators still vulgarly supposed to be peculiarly human and inapplicable to non-human creatures. Most of these concern culture and relate to a claim often made at a high level of analysis or generalization: that culture itself is a uniquely human treasure. Before tackling this claim, it will be useful to look at some peculiar features of human culture which other animals are supposed not to share.

One of the commonest false assertions in this category is that humans are uniquely tool-using or tool-making animals. Tool use is the headline item in the discarded checklist of humanhood. In the forest of Bossou in Guinea, apes in the wild use much the same technology to crack nuts as humans inhabiting the same environment do: two stones—one as an anvil, the other as a hammer; meanwhile, high in the trees, chimpanzees use leaf-stalks to drill into palm-tree pith for nutritious fibre and sap. In the Tai forest in Côte d'Ivoire, chimpanzees wield ten-kilogram stones in a similar way to crack the armour-like carapace of the panda nut, often shaping a small twig to extract the most inaccessible kernels when the main job is done. This is clearly not an innate skill: on the contrary, it takes a young chimp, on average, three years to learn the basic technique, which requires a good deal of delicacy, and five years to master it with real proficiency.

Jane Goodall, the mould-breaking primatologist whose fieldwork uncovered numerous previously unknown aspects of chimpanzee behaviour, showed how even in the wild, without human instruction, some chimpanzees actually manufacture tools—shaping branches to break into termites' nests or chewing leaves to make sponges. Of course, no ape makes tools even remotely as complex as those manufactured by hominid ancestors of modern humans hundreds of thousands of years ago: the world's reputedly 'smartest ape', the bonobo Kanzi, who lives with some of the world's top scientists on a university campus in Atlanta, learned to knap flints for cutting the string with which sweet-packets were bound, but he could never master the particular technique used by *Homo habilis*. Still, apes make the tools they need for their own purposes: the difference between 'them' and 'us' is a matter of enormous degree, but still only of degree.

A stronger and more common assertion about human uniqueness is that only humans have language. Of course, whether this should be regarded as a matter of culture or as an 'instinct', and, if the latter, whether evolution produced it or whether it is some special property inexplicable in evolutionary terms, are themselves matters of intense debate. For our understanding of human nature, the implications of this debate remain a matter of doubt. It is a question of perennial interest to philosophers but, until recently, seemed beyond scientific investigation. The only possible experiments involved isolating children to see 'what language' they came to speak among themselves; though such experiments were occasionally reported in ancient and medieval literature they were extremely rare, because they demanded despotic intervention to get them going, and—though they might have demonstrated, as does the 'private language' sometimes shared by twins before they communicate more widely, that language is in some sense probably innate—they were generally inconclusive: Frederick II's, for instance, in thirteenth-century Sicily, failed, as a chronicler reported, because 'all the children died'.

A more promising phase of the enquiry started when Noam Chomsky became impressed at how quickly and easily children learn speech. He wrote, 'Children learn language from positive evidence only (corrections not being required or relevant), and . . . know the facts without relevant experience in a wide array of complex cases.' They can combine words in ways they have never actually heard (although this may in a sense be an over-rated skill: as we shall see, apes and parrots have it, too). Chomsky also found it remarkable that the differences between languages appear superficial compared with the 'deep structures'—or D-structures, as he later preferred to call them: the parts of speech, the relationships between terms which we call 'grammar' and 'syntax', that are common to all of them. This suggested a link between the structures of language and brain: we learn languages fast because their structure is already part of the way we think. The suggestion was revolutionary when Chomsky made it in 1957, because prevailing orthodoxies—such as Freud's psychiatry and Piaget's educational nostrums—suggested otherwise: that we grow into understanding by developmental stages.

More challengingly still, neither experience nor heredity—if Chomsky is right—makes us the whole of what we are. Part of our nature is hard-wired into our brains. The way Chomsky saw it, at least at first, was that this 'language instinct' or 'language faculty' was untouchable, and therefore perhaps unproduced, by evolution—which really would put humans in a category specially privileged in nature. He has proposed that other kinds of knowledge may turn out to resemble language in these respects:

> that the same is true in other areas where humans are capable of acquiring rich and highly articulated systems of knowledge under the triggering and shaping effects of experience, and it may well be that similar ideas are relevant for the investigation of how we acquire scientific knowledge . . . because of our mental constitution.

At first, Chomsky's claims seem to endorse claims of human uniqueness. We do not know for certain of any other animals with this advantage; we can be reasonably certain that many species have nothing like it. Chomsky, however, was quick to repudiate conclusions favourable to humans' self-image as the climax of creation. On the contrary, our language prowess, on which we tend to congratulate ourselves as a species, is simply another peculiar skill, like the peculiar skills of other species—the bat's radar, for instance, or the spider's ability to extrude a web. Chomsky says:

> It is the richness and specificity of instinct of animals, that accounts for their remarkable achievements in some domains and lack of ability in others, so the argument runs, whereas humans, lacking such . . . instinctual structure, are free to think, speak and discover. . . . Both the logic of the problem and what we are now coming to understand suggest that this is not the correct way to identify the position of humans in the world.

Unless the disposition to language is a special power of the mind, beyond explanation, it must in principle be accessible to more than one species; indeed, our ignorance of the methods non-human species use to express themselves and to communicate makes assertions of human uniqueness in this respect unconvincing. Alternatively, if only humans have language, how might such a situation have come about? No explanation which by-passes evolution can be convincing. And, in search of an evolutionary explanation, the most obvious route lies through the study of the non-human animals most like ourselves. Primatologists have ingeniously imagined how our hominid ancestors might have diverged from our primate cousins and acquired language on the way: speech arose as an alternative to grooming. The growth of the group determined the need for ways of networking that were elastic, inclusive, and time-saving. A look at ape networking today suggests what might have happened.

Ape communities are small by human standards. They tend to disperse. Chimpanzee communities rarely congregate in a single place at one time, while those of bonobos are fluid because females cross community boundaries to mate with perfect ease. Foraging, hunting, and war are all activities in which most apes engage in small groups, rather than fully collective activities in which whole communities unite. The bigger the communities, or the bigger the groups assigned to particular tasks, the more time individuals have to spend networking or cultivating their working relationships. If this has to be done by grooming, the sacrifice of time can simply become too demanding, taking too long, curtailing time for other activities. Species like ours—physically inadept and relatively weak by comparison with many predators—have always had to find strength in numbers and security in collaboration. Robin Dunbar, the leading specialist in the field, reckons that bands of the size normal for hominids a million years ago would have spent 42 per cent of their time on grooming—more than twice that normal for other primates—if they had not devised an alternative socialization technique. Humans have language because they need it. Apes, by this reasoning, do not need it and therefore ought not to have it.

It is true that apes do not seem—to human observers, at least—to have systems of communication capable of attaining the range of human speech; but they communicate for their own puposes, sparingly using vocalizations within the range permitted by their relatively poorly adapted speech organs, supplemented by gesture and grimace. Most communication even among humans is non-verbal and therefore does not depend on the specialized larynx and vocal tract, which make human vocalizations uniquely human. Most of the human repertoire of grimace and gesture is part of a communication system common to primates. Just about all primatologists who have worked closely with great apes acknowledge that they are much superior to humans in their skill at non-verbal communication, reading signals in each other's eyes: a mischievous commentator might liken this to the 'thought-communication' which popular science fiction has often imagined as a higher evolutionary recourse than the mere language with which we humans are crudely equipped today.

There therefore seems little mystery in the fact that human-style language is not part of apes' repertoire of social skills in their own homes. Among humans, however, they learn to exchange language with human companions with startling fluency, using conventional sign language or symbolic code languages punched from computer-style keyboards. Apes have brains that seem well suited to develop human-style language, with areas analogous to Broca's and Wernicke's—the areas most involved in the processing and production of human speech. Indeed, chimps have proved better, on the whole, at learning human language than human researchers have proved in mastering ape communication—an outcome predicted by Montaigne, who thought 'they may as well esteem us beasts as we them.' Dian Fossey made startling progress in exploring the way the mountain 'gorillas in the mist' of Rwanda communicate vocally. She learnt to make the sounds which signify peace, friendliness, reassurance, consolation by breathing stertorously, sucking or blowing through nostrils or teeth in ways that do not sound like speech to a human ear, but which seem to make sense to a gorilla. Her learning experience was cut short by her murder by poachers in 1985.

Meanwhile, Maurice Temerlin raised a chimp called Lucy to 'grow up human'. Lucy chatted using American Sign Language with the level at least of a two-year-old human child, turning the pages of a magazine and interjecting such comments as 'That dog' and 'That blue' at appropriate points. She would take visitors by the hand, stroll with them in the garden, and point out birds and plants with all the pride of a home-owner, signing their names as she went. When her pet kitten died, after a period of touching grief, she came across a picture of herself and her pet in a magazine. She stared at it for a long time, frequently signing, 'Lucy's cat'. One of the all-time outstanding chimp linguists, Washoe, who demonstrated her talents first as a pet, then as a laboratory specimen, in the 1960s and 1970s, invented her own terms for objects the names of which she did not know by combining terms in her lexicon ('rock berry' for brazil nut) and even taught some language to a chimp newly recruited to her lab—a facility which has since become common among apes engaged in learning human language. Any doubt about whether chimps 'really' understood the signs they learned to use was dispersed by the personal tragedy that clouded Washoe's last wretched years of life in captivity. Her sick baby died and was never returned to her arms: from then on, whenever her carer approached her cage Washoe made the same signs: 'Bring baby, bring baby'. Experiments with gorillas and orang-utans—though impressive claims have been made—have not so far been conducted with the same rigour as those with chimps and bonobos; but the results yielded for the two latter species are too consistent to be discounted. They have, unquestionably, a capacity for human language; in suitably contrived environments, they acquire it by imitation from humans, without recourse to Pavlovian training; and when they know it, they use it to themselves and among themselves.

Chimps' knowledge of human language or, at least, their ability to deploy it, does seem circumscribed by insuperable limits. Normally, for a chimp or bonobo, 150–200 terms seem to be an attainable human vocabulary, which cannot be greatly exceeded—but this seems impressive enough across the chasm of millions of years of evolution that separate them from us. No ape has made much progress in mastering the complexities of syntax which seem to come so easily to most human language-learners. Steven Pinker has presented a strong case for the peculiarity of grammar and syntax to humans. However, some bonobos certainly seem able to distinguish shades of meaning conferred by varying word order—as do dolphins—and many ape subjects respond to complex strings of terms and formulate others of their own. It is doubtful, in any case, whether syntactical dexterity should be considered a distinguishing peculiarity of ours, or merely a difference of degree.

Even if the concept of human language were genuinely alien to apes, it would perhaps be more helpful to test its intelligibility for animals who do have vocal structures adapted to make human-style sounds. There is an old debate about whether the vocalizations of birds resemble human language: some investigators have explored this problem among birds capable of imitating speech. The most famous case in history was related by Locke in *An Essay Concerning Human Understanding*. He told of how Prince Maurice of Nassau, one of the most committed and munificent patrons of seventeenth-century science, conducted a remarkable conversation with a parrot while in Brazil as governor of the Dutch conquests there. Giving apparently direct answers to direct questions, the parrot told the prince that he hailed from Maranhão, belonged to a Portuguese, and kept chickens. The authenticity of the story has been doubted, on the grounds that Maurice relied on translators as intermediaries in the exchange; but many parrot-owners have supported it anecdotally with experiences of their own. Now the ornithologist Irene Pepperberg has claimed to have settled Locke's doubts and resolved the long-debated question of whether talking parrots 'understand' what they say: hers answer simple questions with unfailing accuracy and even sometimes, when confronted with unfamiliar objects, apply familiar rules to formulate names. Are her experiments credible to specialists in related fields? The debate goes on, but most of the evidence does seem to be accumulating on one side of it.

Other species-specific communication systems seem analogous to human language, even though they resemble it only very remotely: the dolphin's whistling, the 'dance' of bees, the 'squeaking' of ants, which, according to the claims of a recent researcher, is, among ant-interlocutors, an intelligible code. The animals people work with are often better at distinguishing the sounds made by human voices than, say, Westerners are in detecting the distinctions of pitch by which meaning is conferred in Chinese. In suitably modest degree, the story of Dr Doolittle could come true, but Doolittle is himself likely to be a non-human animal. The best available conclusion in our present state of knowledge is that there are many species with forms of communication specific to themselves, and it is unclear why language—even if it is in some sense a peculiarly human resource—should be treated as a basis for classifying the species that uses it apart from all others.

Language, of course, is a kind of symbolic system, in which words or signs encode the realities they represent. If it were true, the claim that humans were unique in having language would raise a broader presumption about our possible uniqueness as symbol-makers—ritualizing life, associating actions and objects with significance that transcends their palpable effects. Is this, for instance, what makes the practices of macaque monkeys—who, as we shall see, process food by washing it—'protocultural', as some specialists like to say, rather than fully cultural, whereas when humans prepare food it is a rite with meaning?

A representative claim of this type is that only humans have art. The first thing to be said about this is that not all humans do. In 1909, when Walter Grainge White studied the Mawken or oran laut—the destitute 'sea people' of the Bay of Bengal, who had been driven to take refuge on the ocean from earthbound enemies—the deficiency of art was one of the things that most puzzled him. Apart from mats woven patternless, with apparently single-minded concentration on practical utility, they had nothing—no tools carved with patterns or images, no daubings, no dyed garments, not even music or dance. When he asked them why, they replied that they had abandoned them in their 'time of sadness'. This surely does not mean that these people were incapable of art; rather, they chose not to make symbols or images or evocations of things not present. Can the apparent absence of what we recognize as art among non-human animals be of the same sort? Even between human cultures, the differences in the way art is understood and shaped are enormous: we look at each other's works and ask, 'Is it art?' So there should in principle be no reason why we should not look at the behaviour of other species and ask the same question unprejudicially.

Art is the realization of what is imagined (for even a relatively uncontrived photograph or one of Duchamp's *objets trouvés* is changed by appropriation by the artist) and we can be sure that many non-human animals have powers of imagination similar to our own: imagination is a vital mechanism in obtaining food and shelter, reading the weather, anticipating predators and rivals. So potentially, at least, such animals are artists. Chimpanzees clearly understand the idea of art: although, even under human instruction, they do not produce what an adult human's eye normally recognizes as representational art, they do sometimes label their pictures in sign language, saying, 'This apple' or 'This bird' and so on. We do not demand of human artists that they should represent the world as the rest of us see it. Since it would be silly to expect beings of another species to see reality as we do, we should not be surprised if they represent it differently, nor be unwilling to classify those representations as artistic. In at least one respect, some non-human apes do—even without training—manifest a symbolic imagination similar to our own. It is commonly and correctly said that apes never adorn themselves in the wild with the kind of bijouterie favoured by humans—we can look at rats' teeth, say, and see them as items of adornment or social status or magic power, to be drilled and strung and slung round our necks, whereas other apes see only what is immediately, palpably there. However, at the Yerkis Institute of Primatology in Atlanta, female bonobos sometimes put dead rats or cockroaches on their heads and keep them there all day, deriving apparent gratification from the fact. The parallel with behatted human ladies at Ascot is hard to resist. The transformation of dead vermin into items of *haute couture* is not, after all, unintelligible even in human terms, as any wearer of squirrel-skin or fox-fur will be obliged to admit. To see a dead cockroach and re-imagine it as headgear requires a mind capable of inventive transformations.

Use of fire, like art, is a uniquely human achievement which does, however, seem within the potential of some other animals. Apes can learn to light cigarettes or strike flame to release the odour of incense, and even to keep a kindled fire alight. Some humans do not do much better. Some traditional peoples of Australia will not kindle fire but have to borrow it from neighbouring cultures—but whether because of ignorance of the technique or some sacred fastidiousness has never been clarified. The primatological fieldworker Anne Russon reported a case of an orang-utan, re-introduced to the wild after captivity, who worked at fire-making by juxtaposing glowing embers with dry wood, 'blowing and fanning glowing embers with a saucepan lid', and appropriating kerosene to add to the hearth. The *Jungle Book* fantasy of the ape who wants to seize the secret of fire in order to be able to compete with humans turns out to be true. Many animals are attracted to the embers of naturally occurring fires, where they sift for roasted seeds and insects made edible by burning. This behaviour is observable among chimpanzees in the wild and suggests a context for the origins of cooking: to a creature of imagination and dexterity, some of the features of burnt-out woodland, such as the piles of ash and the partly burned trunks of fallen trees, might have appeared as natural ovens, smouldering with manageable heat, in which tough-husked seeds or rough-skinned pulses, unchewable legumes and cartilaginous flesh could be processed.

Yet these considerations may miss the really interesting fact about fire: for it is not only a technology for controlling the environment, and for cooking parts of it, but also a source of socially generative power, which, in human history, has created a focus—'focus' literally means 'hearth'—for socially defining rites. Those rites that bind us most tightly to each other involve the communal cooking and sharing of food. Even without cooking, some chimpanzees seem to have developed or to be developing similar rites. The way some chimpanzee communities share hunted food, even though it is distributed raw, resembles a ritual: the gestures with which food is begged and the order in which it is shared are different from those observed at meals of foraged and gathered foods.

It 'resembles a ritual'. Could it really be a ritual? 'Cooperative hunting and nepotistic food-sharing, Machiavellian social tactics, infanticide, sex for bonding rather than reproduction and intergroup warfare', according to a recent primatologist's list, are among other behaviours, formerly thought to be uniquely human, now known to occur among chimpanzees. Perhaps the most remarkable case of all is that of the chimpanzee 'rain dance' observed at the Yerkis Institute. At the approach of heavy rain, male chimpanzees gather to sway and stamp in a concerted fashion, inducing their keepers to joke about whether their intention is to welcome the rain or ward it off. It looks like a deliberate attempt at a magical practice—evidence of a sense of transcendence or of the power of prayer, such as has often been attributed to animals in anthropomorphic folklore. It is hard to accept such an explanation, but equally hard to think of a better one.

Even if it were true that there are elements of culture that are capable of being used to define humankind, it would not necessarily justify us in hiving our species off from the rest of creation in a special category of our own. Most of the uniquely or typically human features of what we call culture could be evolutionary in origin, and although handling tools and fire, recognizing art or music, and so on are behaviours we learn, our general propensity to adopt those behaviours may be inherited or—for those who like the word—instinctive. In some ways, the culture test has served its advocates as a secular substitute for a religious test—an attempt to find an exclusive criterion of humankind without appealing to the soul or to differentiation by divine intervention. But it has failed because, in any case, non-human animals really do have culture.

Of course, it depends what one means by 'culture'; but if one adopts the most widely accepted definition and acknowledges that culture is any widespread behaviour that is transmitted by learning rather than acquired by inheritance, then some non-human animals certainly have it—and to call it 'protoculture' smacks of evasion. The clearest proof was revealed by observations of macaque monkeys on the island of Koshima by Japanese zoologists in the 1950s: they actually saw a monkey genius—a young female whom her human observers named Imo—introduce two behavioural innovations, which other monkeys copied until they became universal in the group. First, in 1953, Imo discovered a way of washing the dirt from sweet potatoes (with which a local farmer fed the monkeys) by rinsing them in a stream. Her mother, followed by other monkeys who had close ties with Imo, began to do so, too: this shows that the learning spread in a social context. Eventually, only a few old males failed to adopt the practice. Meanwhile, the sea, rather than the stream, became the favoured location for it. To this day, the monkeys still wash their sweet potatoes, even though they are now issued exclusively with dirt-free, shop-bought rations. The custom seems not only to have been transmitted by learning but also to have survived its usefulness—becoming 'pure' culture, like the perpetuation of a rite (although the possibility remains that the practice is still functional, as a means of salting the tubers in brine). In 1956, Imo made another breakthrough: until then, the monkeys enjoyed wheat scattered for them on the beach by human benefactors, but had trouble separating it from clinging sand. Imo flung handfuls of mixed sand and wheat into the water and exploited the fact that the sand sank faster in order to scoop the wheat-grains. This behaviour also spread to the other monkeys. Since then, primatologists have discovered many more such cases all over the world, and culture in much the same sense has been shown—or at least strongly suggested—among elephants, rats, whales, and various kinds of birds.

In communities—human and non-human—that have culture, culture can obviously develop and change. Communities of a single species can become culturally differentiated. A lifetime's study of human history has convinced me that one of the great unsolved—indeed, unbroached—questions about human societies is, 'Why, compared with those of other animals who lead social lives, are they so mutable?' Or, to express a similar question with a distinct comparative emphasis: 'Why have human societies grown so different from each other?' In raising such problems, however, one should bear two caveats in mind.

Firstly, these processes of cultural volatility and mutual cultural differentiation have occupied a relatively brief period of the human past: they really got going only during the latter part of the last great Ice Age. Until then, most human communities had much the same way of life, the same technologies and, as far as we know, other common features of culture, such as the same or similar religions and aesthetics. We simply do not know why the normal continuities of human life were interrupted in favour of the rapid, revolutionary changes which have gathered place ever since: in part, presumably, climatic instability and environmental diversity helped to set change in motion; and change, like appetite, *vient en mangeant*, as changes provoke other changes. It should be remembered, however, that rapid change—viewed in the context of the entire human past—is still an exceptional circumstance. The period of our cultural differentiation (though not of the mutability of our culture) now seems to be over, as globalization imposes, world-wide, a convergent model of how to live.

Secondly, even in the period in question, some human societies have remained largely exempt from change: some forest and desert peoples have attained the stunning achievement of resisting change and conserving tradition with amazing tenacity. If we ask, 'Which have been the world's most successful societes?' we tend to leap to the glib, self-flattering assumption that change is the brand of success; societies that have achieved spectacular progress, expansion, and environmental transformation are hailed as 'great civilizations' and models to copy, even if they have run out of steam or crumbled to ruins. But if survival is the goal, the most successful societies are really those that have changed least—that have preserved their traditions and identities intact, or that have perpetuated their existence by rationally limiting the exploitation of their environments. The longest-enduring societies—those that have successfully resisted the risks of change—are those that still lead the forager's way of life: the Kung San or Bushmen of South Africa, Australian aboriginals, some rarely encountered forest peoples.

This social longevity—as we might call it—aligns them with most non-human social animals. We do know of some instances of cultural change in primate societies—of non-human societies with 'histories' of change. For example, in Tanzania, different groups of chimpanzees have developed different practices—one is tempted to say 'different rites' of mutual grooming. Similar differentiation has also been observed in geographically separated communities of orang-utans (below, p. 56). In Ethiopia, baboon societies in the highlands are characterized by tightly controlled harems herded by individual males, whereas the baboons of the savanna contract much looser relationships of 'serial monogamy'. In Gabon, gorillas in Lopé eat mound-building termites but not weaver-ants; neighbours at Belinga eat weaver-ants but scorn the termites. Different groups of primates of a single species have their own ways of dealing with peculiarities in their environments: using leaves as cushions, for instance, where ground is very moist, or strewing sticks as a form of matting to protect themselves from thorns. These slow, selective changes are, of course, of a different order of magnitude from the 'revolutions' that have repeatedly convulsed some human groups in the last ten thousand years or so; and most non-human animal societies are remarkable for their virtually or utterly unchanging trajectories. Still, even by the criteria of cultural mutability, most human societies, for most of history, have exceeded them in this respect on a scale which, by some standards of comparison, seems only modest.

If we suspend as unprofitable the search for a cultural way of differentiating humans from other animals, is it possible to take another approach and go 'soul-searching'—trying to identify some uniquely human power or quality of mind—whether developed in evolution or conferred by creation, that no other animals have? The quest is of respectable antiquity and has most often led to the privileging of reason. That only humans reason is an assumption well documented in some of the world's most ancient philosophical texts. The uniqueness of the rational soul was Aristotle's defining characteristic for man: other animals had souls that enabled them to act and feel but not to think rationally. Of course, the appeal of such a claim depends on what one means by 'reason': in relevant contexts, at a minimal level of ambition, it generally means a way of thinking that transcends instinct and attains purposiveness. It does not necessarily mean a capacity to think logically or to solve problems in efficient ways: if it did, most humans would have to be excluded from it. Purposiveness defines it, not because of any special quality of purposiveness, but because a creature with awareness of its purpose has something more, which really is special: freedom or, at least, the hope or illusion of freedom. The creature of instinct does not have free will; the creature of reason does or is capable of thinking it does. This is why reason has been so privileged in definitions of humankind traditional in the West. Freedom is a prize with a theological dimension. It is the basis of a further claim which is attractive and may in some sense be true, but which is beyond verification or falsification: the claim that only humans make moral choices.

Instinct itself is in any case hard to define; it seems inadequate to account for all animal behaviour; and, since instinct must generally be supposed to be functional, there is no objective test for an action produced by purposiveness as against one produced by instinct. A precondition of purposiveness is self-awareness, and this in its turn has often been claimed as uniquely human. By the only test so far devised, chimpanzees, bonobos, orang-utans, and some gorillas have it: they recognize themselves in a mirror. If, on parts of their bodies invisible without the use of a mirror, dye is applied while they sleep, they will use the mirror to examine and explore the affected spot. The super-powers on which we have congratulated ourselves are not, therefore, it seems, necessarily confined to the evolutionary paths that lead to humankind. Little is known about the minds of dolphins and whales, but experiments analogous to those with apes suggest they have a cognitive range at least as impressive, by human standards.

It may be possible to define self-awareness and other varieties of cognition in ways which are peculiarly human; indeed, many people expend a great deal of thought attempting to do so. Social anthropologist Michael Tomasello, for instance, has argued very impressively that

> non-human primates have many cognitive skills involving physical objects and events—including an understanding of relational categories and basic antecedent-consequent event

> sequences—but they do not perceive or understand underlying causes as mediating the dynamic relations among these objects and events.

But the difference between causes and 'basic antecedent-consequent event sequences' seems rather problematic, and how, in any case, would one devise a satisfactory test of a subject's understanding of such a difference—in humans, let alone in apes? All claims made for the uniqueness of human cognition are ultimately unimpressive, because similar claims could equally well be made for other species: ascribing to them unique forms of cognition of their own. The result would be that humans would forfeit their claim to unique status. In any case, the claim could not be satisfactorily tested, since we have no means of translating evidence of cognition in a given species into the communication system of any other.

None of this evidence and none of these arguments excludes the possibility that humans may be peculiar creatures in a religious sense: God's Chosen Species, invested, for instance, with immortal souls, or entrusted with divine stewardship or some other special place in God's Providence, or hallowed, as Christians think, by a particular instance of divine incarnation. Those of us, however, who see humankind in this way, and who revere our fellow humans for it, should not be tempted to ascribe other sorts of specialness to ourselves, compared with other animals: that would, I suspect, be to succumb to a form of species-arrogance unintended by a God Who prefers the humble and meek, and Who, if He peculiarly favours us, surely does so in the same spirit of discrimination which made Him prefer a manger to a bed, a fisher to a king, a virgin to a vamp, and a cross to a throne. Nor should our religious reverence—if we have it—for humankind make us love other animals less: on the contrary, if it works to the full, it will inspire and inform our attitudes to them, encouraging responsibility, outlawing cruelty. Because religion and science belong to different spheres of experience and different realms of thought, that should not make them incompatible, but rather complementary; still, because not everyone is religious, it is more useful to look to science than to religion for a universally acceptable concept of what it means to be human.

Now it is hard to speak scientifically about morals because, in a strict sense, scientifically explicable morality is a contradiction in terms. Goodness is not really goodness if it merely confers an evolutionary advantage or some other computable benefit: for it then becomes a form of selfishness. Compassion becomes externalized fear; generosity becomes an investment in feedback; sympathy becomes a collaborative strategy; love becomes—as Diderot put it—'pleasurable throbbing in a pair of intestines'. To be truly altruistic, or truly selfless or self-sacrificial, a moral act must be beyond explanation. Therefore we have to confine ourselves to the most rudimentary kind of science: putting explanation aside and relying on mere observation. All we can do is scrutinize the behaviour of non-human animals for evidence of the same kinds of act which we deem moral in our own species. If we again look, for example, at chimpanzees, we see sympathy and empathy, friendship and disinterested deference, reciprocity and the obligations that come with it, acts of reconciliation and consolation, and even of self-sacrifice all in abundance. Primatological fieldwork reports are full of accounts of fights between two chimpanzees, after which some of the other members of the community will approach the victim and make the standard gesture of consolation by putting an arm around him. Nadie Coates, the Russian primatologist, found she could never successfully use threats or treats to circumvent her pet chimp's recalcitrance, but if she feigned pain he would always return to her to console her, with a look on his face which she interpreted as compassion. Washoe, in a famous incident in her lab, once saved a fellow chimp from drowning and appeared to look reproachfully at human bystanders who had failed to help.

Whether 'animal' morality is human or human morality 'merely animal', the problem of distinguishing the two must be acknowledged as a problem. It seems a fair conclusion to say that—as far as knowledge we can reasonably class as scientific goes—the differences between our species and others are probably of a comparable order, neither much greater nor much less, with those that separate non-human species from each other. Humans are unique, but not with any unique sort of uniqueness.

1. Charles Laughton as Dr Moreau, with one of the quasi-humans he crafted by vivisection, in the first (1932) film version of H. G. Wells's story: 'I took a gorilla … and … thought him a fair specimen of the negroid type when I had finished.' Moreau's experiments failed because 'stubborn beast-flesh' reasserted itself. In the latest movie version (with Marlon Brando, 1996) he uses genetic engineering.

2. Congo the chimp, from *Men and Apes* by Ramona and Desmond Morris. Because many non-human apes and monkeys draw and paint without encouragement or reward, Morris argued that they produce 'art for art's sake'. Congo produced 384 pictures in three years, usually working for about half an hour at any one time before getting bored.

3. The German conceptual artist Rosemarie Trockel features apes with the human habit of trying to capture the world by imitating it. In 'Untitled' (1984), the dancers seem to emerge from the ape's mind as imaginary beings, or aspirations – allusions to a conceit familiar in many cultures: we are the apes who want to be human.

4. To provide subjects for his art, Paul Kane explored the Canadian west for material on the ethnography of native peoples in 1846-8. He was an inveterate romanticizer, who borrowed compositions from Raphael and faces from Reynolds; but he documented costume accurately, including shamanic disguises.

PL. XXIII. — FIGURES ANTHROPOÏDES. — 1. Femme aurignacienne rampant, tracée sur argile (Grotte de Cabrerets, Lot). 2 et 3. Têtes gravées de la caverne de Marsoulas. 4. Le *Sorcier* de la grotte des Trois-Frères. 5. Hornos de la Peña. 6. Altamira. 7. Les Combarelles. 8. Un des diablotins du bâton de Teyjat.

5. One of the clearest early depictions of shamanism appeared on a cave-wall painted some 20,000 years ago, in the depths and grip of the world's last great ice age, at Les Trois Frères cave, Arièges. The context suggests a hunt. The disguised figure faces prey in a head-on relationship which evokes a clash of powers.

6. In numerous repoussé silver plaques of the fourth century BC, from northern Bulgaria, the ancient Thracian Master of Animals displays magical sympathy with and control of animals and monsters.

7. In *Utriusque Cosmi Historia*, his study of supposed correspondences between man and cosmos, the Hermetist Robert Fludd (1574-1638) shows man ascending from ape-like imitation to godlike understanding, dominating prophecy, geomancy, mnemotechnics, horoscope-casting, physiognomy, chiromancy, and pyramidology.

Dumb Friends: Human Solidarity with Other Animals

The inescapable continuities between humans and other animals tempt us to a delightfully ironic suspicion: that our species is unique only in its desire to differentiate itself from others.

Such a suspicion would probably be misleading; self-differentiation from other animals is certainly not an inherent property, instinct, or evolved trait in human behaviour. On the contrary, our insistence on our differentness seems to be a peculiar feature of a late phase of history and of a few unusual cultures. It must be culturally induced, since some societies—and in the past perhaps all societies—have been without it, or have possessed it only in a very attenuated form. Now, we turn to the history of human thought about the animal frontier and how the long and still inconclusive effort to fix it began.

As with all assertions about prehistoric thought, the deficiency of evidence about pre-literate concepts of humankind has to be made up from archaeology and the analogies anthropologists reveal in societies still in touch with their most ancient wisdom. On this basis, our remote ancestors seem to have accepted without question that they were part of the great animal continuum. For the earliest creatures we might reasonably classify as human, this was actually an observable fact, since for most of the hominid past they co-existed with other, similar species. Primitive wisdom, moreover, was bound to defer to animals bigger, stronger, or faster than humans. Animals who were enemies were treated with awe, those who were allies with admiration. We can observe a stirring example among the mesolithic hunters who have left us their graveyard—the largest to survive from their era—at Skateholm on the Baltic. They accepted their dogs as full members of society, burying them with the spoils due to prowess and, in some cases, with more signs of honour, such as gifts of rare, blood-red ochre and spoils of the hunt, than are found in the graves of men. For most of the past, people adopted totemic ancestors, worshipped animals or zoomorphic gods, and clad their shaman-elites in animal disguises, of pelts and feathers, horns and beast-like face-masks. That humans are animals was a truth recognized by most people in most cultures for most of the past: recent history has had to rediscover it.

Totemism is part of the proof. The term has been loosely used to cover any thinking that relates humans and other life forms closely together; in its most powerful form, the totem is a device for re-imagining social ties. In totemism, a relationship with a particular animal (or, in some cases, it may be a plant species or some other natural object, or clusters of a number of such things) defines the group and distinguishes it from other groups who form part of the same society. People's relationship with their totem is usually expressed as common ancestry, sometimes as a form of incarnation. This works because it creates obligations, which transcend self-interest, and constitutes a means of mutual identification. It is a way of keeping track of people of common ancestry in demographically expanding societies—societies of the very sort which peopled most of the world during a spell of about five millennia, when the world's population probably multiplied ten- or fifteenfold, beginning about a hundred thousand years ago. Moreover, although it is strongest and most common as a way of conceptualizing real ties of kinship, the totem is a device that also makes it possible for people who are not tied by real blood relationships to behave towards one another as if they were: in most societies with totemic practices, people can be ascribed to the totemic 'clan' as well as being born into it. Dreams reveal totemic relationships undiscerned by sense of kin.

The totem generates common ritual life. Peoples who class themselves according to totemic affiliations usually share rites of veneration of their special animal, and observe peculiar taboos—especially by refraining from eating their totem, or marrying within the group. One way of estimating the antiquity of a cultural practice is to measure the extent of its occurrence: totemism is one of a relatively small class of customs found in every continent, which suggests a very remote origin, before the dispersal of humankind. It would exceed the evidence to claim that it was universal at some early phase of history, but it does seem to be one of the oldest and most effective ways of forging society, creating and preserving effective bonds between smaller groups in expanding societies, and transforming small bands and families into more numerous groups. The common—and common sense—feature of all the theories is that totemism is related to the difference, which can perhaps be qualified as a 'transition', between two very early categories of thought: 'nature', represented by the totemic objects, and 'culture', understood as the conscious relationships between participants in the group. Probably by chance, the totem-idea also reflects a fact about human origins among pre-human—therefore non-human—ancestors.

In totemic thought, then, humans and animals are aligned in the same communities and the same lineages—rather, in a sense, as they are in modern science. The same sense of community between human and non-human creatures typifies shamanism, which, along with totemism, is a widely diffused and therefore, probably, a profoundly ancient practice. Shamans launch their 'journeys' of ecstatic communication with spirit-worlds by an intoxicating range of methods: drugs, drums, dance, and disguise. The transforming power of magic is mediated or, at least, symbolized, through disguise: a way of bidding for control over nature, a strategy for wresting strength from animals and gods. By donning the disguise of a powerful beast—or of a demon or god, if you believe in such things—you seize its powers. Imitation can be part of an agenda of emulation. Wearing mask or pelt, the shaman seeks to change his own nature: to share that of the beast or divine progenitor. The powers—extraordinary by human standards—of some non-human animals as hunters, trackers, runners, leapers, climbers, fliers, producers of fertile litters, expert scavengers, and predators on man raises the presumption that they have a privileged relationship with the gods, or some other mysterious way of manipulating nature. Therefore, perhaps, animal disguises are among the earliest subjects of art. Mircea Eliade arrived at his theory that shamanism was the world's first universal religion while contemplating the figure of a human dancer disguised as a deer, who appeared on a cave wall painted some 20,000 years ago, in the depths and grip of the world's last great Ice Age, at Les Trois Frères cave, Ariège in the Pyrenees. All interpretation has to be tentative when the evidence is so scanty and remote, but the context of the painting suggests a hunt. The disguised figure faces prey in a head-on relationship which evokes a clash of powers. This, in short, seems to be one of the world's first portraits of a shaman, taking on the form of a beast as a means of magical control. Of comparable antiquity is a painting of a bird-headed man in the cave of Lascaux in the Dordogne. Even earlier—perhaps ten millennia earlier—is a recently discovered painting in Chauvet cave, of a figure who is bison-headed with human legs. The conclusion that palaeolithic painters practised a shamanistic religion seems, indeed, to spring from the cave walls. Herbert Kuhn reported his exploration of the cave art of Tuc d'Audoubert with the Abbé Breuil in confidence that he was confronting evidence of the antiquity of a priesthood whose magic was animal magic: he described the subterranean voyage into prehistory through the waterlogged cave, the emergence from darkness, by lantern-light, of a Stone-Age way of looking at the world. The cave ceiling was so low that the explorers had first to crouch, then lie on the deck of their punt. The rock-vaulting 'scraped the tops of the boat's sides. . . . Suddenly, there they were. Pictures. Beasts engraved in the stones . . . Shamans, too: men wearing beast-masks, uncanny figures and weird.'

Magical practices of the same kind are so widely diffused in traditional societies of the present and recent past that they must be a deeply embedded ingredient of culture. Convincing interpretations of historic rock paintings of South African Bushmen, which resemble those of cave art, suggest that they record shamanic rain-making rituals. Disguise is not mimicry, of the kind a parrot can perform; nor is it, like the self-transformation of the chameleon or the spider-crab, or the camouflage of the stick-insect or ermine, a merely 'evolutionary' mechanism to conceal the predator or decoy the prey. Self-consciousness has to precede conscious self-transformation. The notion that virtues are transferable by magic is accessible only to an imagination informed by experience but unrestrained by it.

There are, of course, methods of self-transformation that go deeper than disguise. The earliest evidence we have of body-sculpting is in the deformed crania of a whole generation of people buried some 12,000 years ago in the deepest stratigraphic layer investigated by archaeologists at Jericho. Anthropological evidence suggests that it is often associated with the imitation of other species. Scarification, for instance, resembles a hide or pelt; in combination with tattooing it can be used to give human skin a herpetic texture and appearance, or echo the patterns of scales. Chiselled teeth acknowledge the superiority of the serrated bite of powerful carnivores. Genital mutilation, in parts of Australia, is tribute to the penis of the kangaroo. Man is indeed the ape of nature. Much body-painting and sculpting—most of it, probably, in the modern West—is intended for cosmetic effect, but even cosmetic effects are often themselves tributes to the beauty or superiority of the animals they imitate: adding to human eyes a lynx-like lining or to human hair the gloss of a pinguid pelt.

Superior Person: Looking Down on Creation

If totemism, zoomorphic religion, and shamanic rites all argue for the antiquity of human sympathy with other animals, the problem arises of when humans' sense of apartness from other creatures arose, and when, in particular, it first occurred to people to claim superiority—the gift of Eden, the lordship or, at least, the stewardship of creation. In view of human feebleness and inferiority, in physical prowess and direct combat, to so many of our competitor species, it seems a counter-intuitive claim. Early heroes of many civilizations tended to be credited with extraordinary powers of magical sympathy with animals, which sometimes involved power of domination. Arguments, for instance, that Indo-European heroes included a 'Master of Animals' rest on Thracian images of a horseborne hunter—lord and exploiter of the natural world: plaques depicting him lie defaced and trampled at early Christian shrines, leaving no doubt of his divine status. In surviving examples we see him taming horses, wrestling monsters, and wresting quarry from a lion's maw.

But the claim that humankind as a whole inherited the supremacy of nature cannot be traced back very far in explicit evidence, not beyond a period well into the first millennium BC—the 'axial age', when so much of our modern thinking was initiated or anticipated by sages in China, India, the Near East, and Greece. The first evidence of a shift in relevant sensibilities occurs—like most of the 'firsts' of history in that period—in India, in the Upanishads. Those revolutionary texts developed even older traditions of thought about human nature, discernible in the Vedas, in which humankind is closely associated with rational qualities. The Vedic word for humankind—*manusya*—is related to *manyas* (mind) and *man* (thought), from which, according to the commentary called *The Brahman of a Hundred Paths,* humans were created. The most famous of Vedic creation hymns puts it differently, but still with a sense of peculiar human refinement: the Creator's limbs were the substance from which humans were fashioned, whereas other animals were made from a kind of chaos of 'milk and ghee'. If this suggests a hierarchical model of creation, with humans ranked higher than other creatures, the Upanishads add an essential difference that puts humans not just into a higher rank, but into a new category of being: a soul, or *atman*, which is not part of nature at all, but is eternal and immutable. This was a radical claim, transforming humans from the elite of nature to a place in supernature. It anticipated future thinking in all other sedentary civilizations. The project of prising people from nature had begun.

In China and the West, the tradition has been deeply affected by the notion that in being elevated above nature, humans acquire dominion or, at least, trusteeship over it. In the first chapter of Genesis, God makes 'man in His own image' as the last word in creation and gives him dominion over other animals. His first commandment is, 'Fill the Earth and subdue it. Be masters of the fish of the sea, the birds of heaven, and all the living creatures that move on Earth.' Though formulated as a divine insight, this was also an acknowledgement of an increasingly obvious fact: people's power to transform their environment and to exploit other species was the most conspicuous common feature of agrarian societies. It is not surprising, therefore, that early in the second half of the millennium, thinkers in other traditions formulated similar ideas. In perhaps the late fifth century BC the tradition ascribed to Mo Ti took shape; that 'everything is prepared for human benefit', including all other creatures, was one of its principles. In the mid-fourth century BC in Greece Aristotle schematized a hierarchy of living souls, in which man's was adjudged superior to those of plants and animals, because it had 'rational', as well as 'vegetative' and 'sensitive', faculties. One Chinese formula, as expressed, for example, by Hs'un Tzu early in the next century was similar to that of Aristotle: 'man has spirits, life, and perception, and in addition the sense of justice; therefore he is the noblest of earthly beings.' Chung-shu in the first century AD explained that humankind was a different order of being from the rest of creation—infused with life 'in a manner markedly different from the great mass of living creatures'. The proof was practical human superiority—exploitation of plants and livestock for food, and the power 'to tame the bull, ride the horse, trap the panther, pen the tiger. It is thus that he has gained spirituality above that of other creatures.'

Buddhism, meanwhile, developed a distinctive doctrine, in which the real discontinuity arose not between humans and animals but between Buddhas and others. Humans shared the base cravings and desires of other creatures—the self-interests, the desire for life, the at-best obnubilated sense of transcendence—and could only realize their potential for elevation above the grossness of the world by rigorous virtue and progressive ascent through successive reincarnations. Still, that potential was confined to humans. Animals and demons started in realms too remote from spirituality for early elevation. For a human to be reincarnate as an animal was demotion—likened in some early Buddhist texts to falling into a cess pit. So, despite its critical self-contemplation, the Buddhist view of humankind was also affected by the apparently inescapable tendency of axial-age thought to tug humankind free of the rest of creation.

In this atmosphere, the common approach of sages all over Eurasia was to define humankind in terms of a check-list of qualities. For Mencius, the main question about humankind was moral: is human nature good, bad, or neutral? His check-list was determined by the terms of that question and by the answer he wanted to give, which was that people naturally tend to be good. 'Commiseration is essential to humankind', he wrote, adding that shame, hatred, modesty, complaisance, and discrimination were all defining characteristics of human beings. Earlier ages would not have assumed that these qualities were inaccessible to non-human animals.

Check-lists, however, are never intellectually satisfactory: the mind craves an overall characterization. By the second century AD in China, the idea of 'man as microcosm' was being adduced as evidence of the supposedly special place of humankind in creation: the uniqueness of humans could be seen in their manifest perfection, their fitness for the mastery of nature in the way their bodies and minds echoed, mirrored, or encompassed nature. Tung Chung-shu took the idea to extraordinary lengths. Man had 'twelve large joints' because the year had twelve months, and 366 lesser joints because the year supposedly had 366 days. There were five viscera because there were five elements. Eyes and ears corresponded to sun and moon, veins to rivers, breath to wind, hair to the stars. The head was round to match the shape of heaven, and the alternations of sadness and pleasure reflected the cosmic balance of yin and yang.

At one level, this was a sort of secular version of the claim that God made humans 'in his own image'; rather than the image of God, the human was the image of nature—which, in some traditions, amounted to the same thing. Through Roman and Hellenistic texts, it became an imbedded notion in Western thinking, too, shaping Renaissance artists' obsession with perfecting the human form and projecting its dimensions onto the universe. Tung Chung-shu anticipated the cry of Miranda, 'Oh, what a thing is Man!'.

The Obstinate Debate: Animal Apologists and Critics of Humankind

But did human superiority mean human privilege or human responsibility: lordship or stewardship? The assertion of humans' special status inaugurated a long, still unresolved debate over the limits of humans' freedom to exploit other creatures for our benefit. From the first, there were dissenting traditions. In southern Italy in the late sixth century BC, Pythagoras taught that 'all things that are born with life in them ought to be treated as kindred'. His Indian near-contemporary, Mahavira Vardamana, thought the universality of souls imposed on people an obligation of care of the whole of the rest of creation: because the souls of animals most closely resembled those of people, they had to be treated with special respect.

More was at stake, however, than the limits of human treatment or mistreatment of supposedly inferior species. Resistance to the very notion of human uniqueness never quite faded away. While other people's gods got progressively anthropomorphic, Egyptian civilization clung to gods with the faces of crocodiles and dogs. Confucius tended to class animals, including man, in one category, and plants in another. Wang Ch'ung (AD 27–97) admitted that among the 300 or 360 'naked creatures' known to the science of his time, humans were 'the noblest and most intelligent'. But it was a matter of degree: in essence, they were still animals, crawling on the Earth 'like lice in the folds of a garment'. If fleas were to speak in men's ears, men would not hear them. Similarly, it was ludicrously presumptuous to suppose that Heaven and Earth would pay heed to man. 'Man', he argued,

> is a creature among the myriad creatures who possess knowledge. . . . Birds have their nests and perches, animals have dens and lairs, fish and scaly creatures their particular habitats, just as men have their homes and mansions. . . . Man lives and dies, and creatures too have their beginnings and ends. . . . In blood, veins, head, eyes, ears and mouth, they are in no way different from man. Only in likes and dislikes is there a difference . . . They share the same heaven and the same Earth and equally gaze up at the same sun and moon.

For Wang Ch'ung, humans were what they appear to scientistic thinkers in our own time: purposeless—cosmic accidents engendered by the hazards of nature, not by a divine plan (though Wang Ch'ung differed from today's materialists by attributing human origins to spontaneous generation rather than evolution).

Taoism, meanwhile, challenged notions of human superiority from another, more spiritual perspective. According to the third-century Taoist compilation known as *Lieh-tzu*, animals and men anciently walked together—as in the Western myth of the golden age—and in a Utopian realm men could still understand animal language. When the ancient master who gave the book its name rhapsodized about heaven's partiality for humankind—on the grounds that grain, fish, and flesh were provided for people to feast on—a disarmingly candid child at the banquet remarked that one might as well say heaven had favoured wolves and tigers by providing them with men to make a meal of. Taoist appreciation of nature kept resurfacing in later Chinese tradition, tugging at Confucian consciences. Ch'eng Hao, for instance, in the eleventh century, acknowledged that 'Man is not the only intelligent creature in the universe. The human mind is the same as that of plants and trees, birds and animals.' And the image of 'Man the microcosm' always had its critics, even at the height of the Western Renaissance: when Shakespeare made Miranda summarize the tradition of Man's perfection, he was surely uttering a similar satire. Miranda's inspiration is not detached philosophical reflection, but impassioned sexual desire. She does not look around for evidence of human distinctiveness from the rest of creation, but straight ahead towards an alluring youth.

Indeed, the debate over whether humans are really 'different' has never ceased, though for most of the intervening centuries, in all the major civilizations of Eurasia, the preponderant view has favoured a special status for humans. It is hard to resist the impression that this is a view recommended largely by self-interested considerations, which themselves undermine the notion that humans are peculiarly moral creatures with uniquely lofty sensibilities. It is obviously a way of legitimating the exploitation of other species. Even during its era of preponderance, people seemed to withhold complete assent; even cultures most inclined to privilege their own species continued to treat other animals as inferior but analogous, crediting them with moral responsibility (and granting them corresponding rights).

In medieval and early modern Europe, for instance, homicidal animals were tried before execution, usually by the same means as prescribed for human murderers. Animals in medieval Europe were acknowledged to be capable of a religious sense and even of attaining sanctity. St Guinefort was a dog: the fastidious preacher Etienne de Bourbon came upon her cult in Villars, near Lyon, in the thirteenth century. She had saved a baby from a snake, then settled down, wounded and bloody-jawed, to watch the crib; seeing her, the child's nurse uttered a scream, whereupon an armed guard rushed in and, misconstruing the scene, slew the dog. Etienne sensed superstition and suspected a pagan survival. He dismantled the shrine, re-buried the dog, and felled and burned the trees planted by local piety, but he could not eradicate the cult, which transferred to a new shrine nearby and continued to worship the canine martyr, while the church looked on, unsuspectingly mistaking the cult for veneration of a Pavian virgin of the same name and impeccable credentials. In medieval Wales, too, pilgrims visited the shrines of canonized dogs: there could be no more powerful demonstration of the moral equivalence of man and beast. St Francis preached to ravens. St Antony of Padua gave communion to his horse. Thomas Hardy's famous Christmas poem recapitulates a genuinely popular English rural tradition about the grace vouchsafed to animals on Christmas Eve, when, at the moment of the anniversary of the Incarnation, they kneel to worship their creator who—with an aptly zoomorphic gesture—was born in a stable: 'Now they are all on their knees. . . .'

Until about three hundred years ago it was still common for animals to have legal rights practically on a par with humans. Rats who despoiled barns, insects who ravaged crops, birds who defecated over shrines, and dogs who bit people were tried in court for their 'crimes', represented by counsel at the community's expense, and were sometimes acquitted. As late as 1650, a horde of grasshoppers was put on trial near Segovia, after the clergy had failed to disperse it with relics, processions, holy water, and public displays of penitence. The witnesses included the patron saints of afflicted villages, represented by villagers, and the judges were saints for whom a friar acted as interlocutor. The rather unorthodox sentence, pronounced by the Virgin Mary, was that the insects would be excommunicated if they did not depart. The prior of the Jeronimite house where the trial took place explained that there were precedents: excommunication had been the punishment of locusts in Avila, rats in Osma, and swallows who fouled a shrine outside Cordova. In London in 1679 a woman and a dog were put to death together for the bestiality they had joined to commit.

Trials of rats, mice, and locusts were also common in China in what we think of as the Middle Ages and early modern period; Chinese folk stories credited non-human animals with human emotions. In tales of Buddhist inspiration, man-killing tigers are often touchingly depicted as remorseful. Of course, this implies the tiger is in some sense a lower order of being than the man—his piety would have little point if it were not an example to his betters. The *Jatakas*—the early stories, influential throughout the Buddhist world, of ascents to Buddhahood—include many tales of animal kindness: a hare who sacrificed himself to feed a Brahmin, shaking his coat before death to detach his lice and spare their lives; an elephant who pardoned the hunter who killed him and gave him his tusk; a snake who lived as a monk; and a peaceful trio of monkey, jackal, and otter who dwelt together 'obeying the moral law'.

The deeper one looks into the history of human convictions of superiority over the rest of creation, the more surprising they appear. Only two forms of physical prowess have been claimed as areas of human specialization. Firstly, we do seem, on average, to be relatively good at throwing, even compared with other primates. This skill has generated our long history as users of missile weapons and originated, presumably, in what would otherwise have been our ancestors' inadequacy in close-up combat. Secondly, we seem to sweat relatively profusely. This is another matter of relatively small degree, an advantage widely shared among mammals, who have developed sweat glands over a period of perhaps two hundred million years. It is in origin, presumably, a function of our ancestors' relatively slow locomotion, which demanded of them extra stamina in the chase: a slower animal can only catch a faster one by keeping going for longer. As they would rarely be able to catch most unwounded prey, humans would need to be able to keep up their pursuit of it with exceptional tenacity. Most of the attributes on which we humans congratulate ourselves seem, in short, to be evolutionary compensations for physical feebleness. Well-developed brainpower is a competitive advantage in a struggle for survival with more powerful rival predators. Tool-making suits species under-equipped in tooth and claw. Language is useful for creatures compelled to huddle in large groups for security and for hairless apes who need a substitute for grooming. Cooking is the obvious recourse for a species short of ruminative skills and reliant on foods which bipedalism makes hard to digest. Against this background, the struggle for human self-definition has been understandably hard and long.

CHAPTER 2

FORMALLY HUMAN

Delineating the Human Body

The story of how the struggle for a definition of humankind was resolved is a long one, characterized at almost every turn by the stunning irrationality of the choices made and the inadequacy of the science invoked to justify them. The ruling principles of what we might call 'humanhood' included possession of reason—but what was reason?—or of a soul—but how do you spot a soul?—or of a body like the observer's—but what differences mattered, and how much? How could you tell humans from the *similitudines hominis* and monstrous beings who filled the nether links of the chain of being? The question was all the more difficult because some of the postulated categories had a habit of disguising themselves as human, or transgressing the boundaries in puzzling ways: the werewolves, succubi, beast-men, and progeny of bestiality. Two avenues of approach led from these problems towards possible solutions: the first, which is the subject of this chapter, focused on an attempt to delineate a definably human body; the second (which forms the story of the next chapter) on a similar attempt to delineate definably human culture.

Excluding the Ape

In January 2003, press reports revealed good news for orang-utans. According to *Science* magazine, they are 'almost human'. Some of them 'use napkins when eating' and 'kiss each other good night'. Some 'use leaves for gloves when handling prickly vegetation', while others wield leafy branches 'as parasols to protect them from the sun'. Even more surprisingly, they develop culture: like human societies, orang-utan groups develop distinctive ways of behaving towards each other. Their games vary from place to place. In Borneo they play by knocking over dead trees, which they bestride as they fall and vacate just before impact. This game, however, is unknown to the orang-utans of Sumatra.

Such findings are part of a current revolution in our understanding of the nature of humankind and of traditional assumptions about the relationships between humans and other animals. In particular, they belong in the context of our growing body of knowledge about non-human culture. Almost every new fact, in what has become a formidable sequence of revelations on this subject, makes us more doubtful of our uniqueness. The orang-utans of Borneo and Sumatra resemble *'Homo ludens'*—social humans whose interactions include play. From a purely primatological perspective—even before one considers the implications for the study of man—this is a remarkable revelation, because orang-utans, who forage alone, have traditionally been classed as 'unsocial apes'. Their societies, like human societies, exhibit potential for development and mutual self-differentiation: indeed, albeit to a modest extent by human standards, they do develop and they do become different from one another. In these respects, the creatures justify the name they are known by, which their Malay neighbours gave them uncounted years ago: 'orang-utan' means 'man of the woods'.

These discoveries about orang-utans have come more than two hundred years too late for their great apologist, James Burnett, Lord Monboddo, who died in 1799, resentfully failing to convince the world that these apes were really ill-classified humans. His theory of social development, gradually elevating man from a beastly condition, was similar—and perhaps contributory—to Darwin's. Like his successor, he sensed people's affinity with other primates. In work he began to publish in 1773, he argued that orang-utans used tools, recognized human music, and could be taught to play the flute. In satirical tribute to Monboddo's views, Thomas Love Peacock—the funniest of England's many great comic novelists—wrote *Melincourt*, published in 1818. The hero is an orang-utan. His name, Sir 'Oran Haut-Ton', plays on the conceit of the ape-as-flautist, while also hinting at the creature's potential politesse. Because Sir Oran had every rational faculty, except the use of speech, he acquired a reputation as 'a profound but cautious thinker'. The consequences included his election to parliament and elevation to a baronetcy. In a sense, modern primatology has worked its way back to a similar conclusion. 'These apes,' ruminated the primatologist Adriaan Kortland in 1973, as he watched Washoe talking to herself in sign language while she leafed through a magazine, 'have a lot more to think than to say.'

Monboddo's sympathy with the orang-utan was, in part, the product of a personal project. His real aim was political: to argue that the state was a contrivance, that men had no 'instinct' or natural disposition to form states, and that Aristotle was wrong to claim that 'man is by nature a political and social animal'. Believing that language was a necessary pre-condition for political association, Monboddo hoped to convince the world that language (as opposed to the 'inarticulate cries' he attributed to early man) is not 'natural' but rather a result of what we would now call culture. As we have already seen, debate on this issue remains open-ended and of great importance for understanding what it means to be human. The basis of Monboddo's argument was philosophical—language is an idea and ideas are not instinctive, but derived from experience or inspiration. He supported it with an empirical claim: 'whole nations have been found without the use of speech'. The humanity of the language-less orang-utan was part of his proof. He first saw one, which had once been a domestic pet, stuffed and mounted in the royal Cabinet of Curiosities in Paris, where he learned that in life the creature 'had as much of the understanding of a man as could be expected from his education, and performed many little offices to the lady with whom he lived'.

Underlying Monboddo's work was a long history of human doubt about the relative place of apes and people in the panorama of life. In embracing the orang-utan, Monboddo was in the vanguard—at the 'cutting-edge', as researchers now say, or even, perhaps, a little beyond it—of the science of the Enlightenment. Part of his reply to his critics was that he was following Linnaeus, who, in 1736, proposed a scheme of classification of all creation, with man classed alongside other apes in the 'order' he called that of 'primates'. Ambiguously, the orang-utan straddled Linnaeus's categories of ape and man: he partnered man in sharing the highest designation, that of 'homo', while still belonging, 'much as this species resembles mankind, . . . to the genus of Ape'. In 1768 the president of the Prussian Academy of Sciences expressed a desire to converse with orang-utans, whose 'natural' wisdom he expected to find more improving than that of his fellow savants. Rousseau—surprisingly, perhaps—was of a somewhat different opinion. He also regarded orang-utans, along with gorillas, of whom he knew almost nothing, as human; but, for him, they belonged to an earlier phase of prehistory than the 'good savage' and demonstrated the lonely, ungregarious state of natural man. Of course, he was wrong: all primates live social lives. But his willingness to classify some apes as human was not unreasonable by the lights of his time.

This educated interest in enlarging the category of humankind was not simply the product of ignorance. On the contrary, the claim that apes are humans was a serious scientific proposition, founded in growing knowledge. The early modern period—the age of the Scientific Revolution and the Enlightenment—was a time of growing familiarity, for Western European savants, with animal relatives who had previously been little-known in the West, but whom the global range of scientific exploration and imperialism now reached. Like many family reunions between long-separated kin, it provoked anguish, raised questions, and revised perceptions. When Richard Jobson, an English trader known only from his writings, in which he exhibits his remarkable powers as an observer, ascended the Gambia in 1620, he was impressed by some of the natives, whose level of civilization he likened to that of the Irish, but by none—in some ways—more than the baboons. Perhaps confusing them with chimpanzees, he claimed that they made clearings for 'dancings and recreations' which resembled 'the handy worke of man'. He recognized them as social creatures who practised 'government, with leaders whose barking could command silence'. Citing the opinion of an unnamed Spaniard, he was inclined to classify them as human beings—'absolutely a race and kind of people'—who voluntarily espoused a 'natural' way of life, avoiding work in order to preserve their liberty and shunning speech in order to preserve their innocence: this last conceit was a widespread topos and recalls the Chinese fable of the 'wise monkeys' who, thanks to their own prudent precautions, hear, see, and speak no evil. Biologically speaking, baboons are fairly remote relations of humankind, but Jobson's insight, which was typical of his time, was not inappropriate: it reflects the resemblances we might expect to strike an unprejudiced observer. And modern science has, obliquely, endorsed it: because baboons are savanna creatures, like the supposed hominid ancestors of modern humans, they are favoured by some anthropologists in the quest for models of early human behaviour.

Shortly after Jobson's encounter, the first specimens of great apes began to reach Europe. This was a great episode in the history of the development of our concept of humankind. For, if baboons could suggest self-reflections for humanity, how much more so could the great apes, whose human resemblances were so much more marked! The first arrival was an orang-utan whom Nicolaas Tulp—the demonstrator immortalized in Rembrandt's *Anatomy Lesson*—illustrated with some accuracy in a book on anatomy in 1641. The creature is touchingly represented, seated (so as to dodge the problem of whether he was truly bipedal) with modestly downcast eyes, nervously twiddling his thumbs; but his patchy shagginess, clawlike feet, and correctly depicted face suggest that this intrusion of sentiment was an honest response of the illustrator to his subject, rather than an attempt to exaggerate the ape's humanity. More influential, among the earliest representations of great apes to reach Europe, was the image transmitted in Jakob Bontius's works on the flora and fauna of the East Indies in the mid-seventeenth century. This greatly exaggerated the human resemblance of the orang-utan: with an upright posture, human face, near-human proportions, and a sparse coat of tufts. This misimpression was laboriously corrected in 1699 after a dissection by Edward Tyson; unfortunately, what the anatomist took to be an orang-utan was evidently, however, a chimpanzee, and Tyson's caution against classing it as human was widely overlooked.

In the eighteenth century, the imaginative appeal of anthropoid apes was irresistible. Singeries decorated walls in aristocratic houses. Teniers, Watteau, Chardin, and Goya amused themselves by depicting apes as artists—playing on the conceit of art as 'the ape of nature'. Acquaintance with apes subverted a traditional form of human self-confidence. For, while the science of anatomy encouraged materialism, demonstrating the continuities between humans and other animals, and disclosing no soul, the apes undermined convictions of human peculiarity and privilege. Gradually or fitfully, the process has continued ever since. The more people see of primates, the more struck they are by their human resemblances and the more obvious it seems that people are part of the great animal continuum.

Indeed, it is tempting to suppose that the Western tradition of radically separating 'animals' and 'men' as discrete categories of beings arose from ignorance of our closest non-human relatives: in parts of the world where apes and monkeys are common, such radical assumptions are less prevalent. Nevertheless, Europeans in antiquity and the Middle Ages knew enough about apes to puzzle over what they knew and to want to know more. The great verse-encyclopaedist of the twelfth century was the ironically named Bernardus Silvestris—ironically, because 'silvestris' literally means 'of the woods', so he might have been expected to sympathize with the category of 'wild men of the woods', in which apes were commonly placed. He regarded them as God's last, botched attempt at creation before the first man—a view remarkably prescient of evolutionary theory. At the same time, he managed to align himself with the usual contradictory view that apes were not men's ancestors but their descendants: 'men of degenerating nature'.

This sort of relationship between man and ape, which pre-inverts the evolutionary model familiar to us, underpins legends of metamorphosis common in many cultures: humans turn into simians in consequence of some transgression. In many cases, this is said to be how apes or monkeys originate. In the *Popol Vuh*, the ancient epic of the Maya, the hero-twins repaid the mockery of their half-brothers, Hunbatz and Hunchouen, by turning them into monkeys: 'and they left the loincloths trailing and these turned into tails. . . . Afterwards, they kept to the treetops, . . . now howling, now falling silent.' Yet the new status of the victims of the trick was by no means contemptible. Though disqualified for rulership they filled high-status, cerebrally demanding occupations as scribes, craftsmen, and musicians. 'They became,' the *Popol Vuh* declares, 'flautists, singers and writers, carvers, jewellers and metalworkers.' Monkeys, indeed, are often depicted discharging such functions in Meso-American art. A Maya vase now in the New Orleans Museum of Art shows the monkey-twins of the *Popol Vuh*, surrounded by books bound in jaguar-skins, eagerly reciting some of the texts—perhaps, indeed, the text of their own magical transformation. 'So they were worshipped,' adds the surviving version of the *Popol Vuh*, 'by the flute-players and singers of old and the writers and sculptors prayed to them.' In Jewish legends the most presumptuous builders of the Tower of Babel were demoted to the rank of apes. Many African myths represent monkeys as the offspring of outcasts from village communities. Zeus turned the mischievous Cercopes into apes in punishment for their fondness for tricks. Sometimes, transformations of this sort happen without any punitive pretext. Baldaeus, the Dutch ethnographer of the seventeenth century, relates the Indian story of Ixora and Paramesceri, who adopted the forms of apes for the conception of their offspring, Sri Hanuman, 'leaping and jumping around the forest until they settled on a bamboo-plant'. The great originality of Charles Darwin, therefore, was not to postulate a close relationship between simians and humans, but to reverse the usual order of events. In his narrative of the descent of man, humans were evolved apes. Previous narratives had represented simians as degenerate humans.

Either way, the difficulty of drawing the line between these contiguous categories arose in the same ways: in part, from physical resemblances between apes and, in part, from continuities of behaviour between humans and other social animals. For there are two possible, potentially conflicting approaches to the problem of defining humankind: deeming it a problem of biology, and seeing it as a problem of culture. According to the former approach, which is the subject of the next few pages, humankind has a distinctive look, shape, or form, or peculiar physical properties. In the latter (which the next chapter addresses) human is as human does, and special ways of life qualify groups for membership of the human community.

Accommodating the Monstrous

Until the Enlightenment era, when scientific taxonomy began to reshuffle the categories in which Western minds arrayed life forms, the prevailing images in the West were of a ladder or chain: the 'ladder of creation', the 'chain of being'. Angels were at the top, closest to God, with people just below them. Brute beasts occupied the nether rungs or links, arranged in order of their resemblance to men, with apes in the upper range of the sequence. But, in the interstices between humans and apes, there was plenty of space to locate speculative or imaginary creatures: *similitudines hominis* (as they came to be called from the thirteenth century), beast-men, monsters with human resemblances, or examples of degeneracy.

The sciapods, one-legged creatures who each reclined beneath the shade of one enormous foot, lurked in the margins of medieval manuscripts, alongside the cynocephali (dog-headed men), tailed men, and those 'men whose heads do grow beneath their shoulders', who were known to Othello and to the imaginations of many travellers of his era. Marco Polo reported the existence in the depths of Asia of specimens of a species which resembled men except for their lack of mouths: they took nourishment by inhaling. Giants were eagerly sought by travellers in unknown lands: the insistence of some voyagers that they had found outsize specimens in Patagonia—the very name means 'Land of Big Feet'—was maintained with startling tenacity. It started in the sixteenth century, when rumours of the presence of giants began with the discovery of supposedly huge footprints in the soil, and lasted until the eighteenth, when reports of expeditions regularly included allusions to natives of absurdly exaggerated stature: a 'pretty Patagonian,' according to the English explorer John Byron, who passed that way in the 1760s, was 'not seven and a half feet high'. By then, however, another agenda had begun to wrench reportage back towards realism, as the science of the

Enlightenment focused seriously on 'the dispute of the New World': the problem of whether the environment of the Americas was naturally stunting and impoverishing, as some leading European thinkers maintained, or whether it nourished improved species, as some creole scientists claimed.

Meanwhile, in a parallel case in Africa, pygmies were known by report and their humanity much debated. Classical Greece received reports of the existence of diminutive people in southern Africa: pictures of them occur on black-and-red urns. Their existence was confirmed by what may have been eye-witness accounts from Portuguese slavers and *degredados* or exiled fugitives who pentrated deep into the Kongo basin in the early sixteenth century, where the Mbuti pygmies still inhabit the Ituri forest. News of them influenced an image which appears in Sebastian Cabot's world map of 1540, where the artist seems anxious to emphasize the pygmies' human status: they are shown in classical poses, conducting what looks like a Renaissance conversazione, handling staves which are presumably included as evidence of their tool-use and dexterity. For Milton, the 'pygmaean race' seemed mythical—like 'fairy elves'—but those who believed in them classified them as human: both their existence and their essentially human nature had, after all, the backing of Aristotle. The inclusive attitude—albeit in a distinctly patronizing form—was maintained by later European explorers, who were able to supply incontrovertible evidence of the reality of pygmy peoples. Sir Henry Morton Stanley's impression of a gazelle-eyed 'little demoiselle' of 33 inches, whom he encountered during his trail-blazing on King Leopold's behalf in the 1880s, was characteristic. It was not beholders from afar who doubted pygmies' humanity: understandably, perhaps, it was their Lemdu neighbours who were most conscious of their otherness, and who hunted them for food with the same moral indifference as chimpanzee hunters of baboons or colobus monkeys: from our distance, the pygmies' alterity does not seem very marked; for those looking close-up at it, it looks like a gulf. Debate about the Congolese pygmies' place in the human panorama is unresolved even today. As I write, in May 2003, press reports from Congo claim that pygmies have issued desperate pleas to the United Nations to save them from the depredations of cannibal neighbours, who hunt them with impunity because 'both government and rebel forces in the war-torn country consider pygmies to be a sub-human species'.

Creatures marginal or abnormal have always tended to congregate in the midst of maps and beyond the edges of civilizations, or in the underexplored recesses of forest, bog, and mountain, such as the tilled lands of sedentary, urban cultures always enclose. In antiquity and the Middle Ages, for instance, the frontier of what had been the Roman Empire and became Latin Christendom was haunted by human simulacra: ogres and elves, vampires and werewolves, wodehouses and willies. This suggests that the Comte de Gobineau—notorious as one of the founders of 'scientific' racism in the nineteenth century—had some grounds for

8. This complex caricature of 1861 likens Darwin not only to an ape but also to Boyet, the smooth-tongued fixer whose match-making efforts fail in *Love's Labour's Lost*. In holding up the mirror to the ape, Darwin is made to echo a traditional gesture of vanity. (left)

9. In the original production of The Magic Flute in 1791 the wild creatures tamed by Tamino's piping included the ape-servants of the evil, priapic blackamoor, Monostatos, who tried to force his attentions on Pamina. (below)

10. The ape as artist was a common conceit for eighteenth-century painters, represented here in a work of 1740 by Chardin. The artist 'aped' nature: the joke, which must have worn thin by the time of this painting, was that the artistic ape was unexemplified in nature and arose from the artist's imagination. Apes' genuine artistic proclivities (see p. 62 and plate 2) were unknown at the time.

11. Aztec depictions of the divine monkey, Ozomatli, often evoke eructation, perhaps because he is a natural musician: here wind- and reed-glyphs emerge from his face. He also commonly displays expertise in dancing and visual and literary arts. He is associated with death – arising here from the aquatic underworld – and sometimes shown symbiotically with the bloodstained skeleton-god, Mictlantecutli.

12. In the Ramayana, Hanuman, king of monkeys, was learned in science, arts, and languages; but his greatest skills were his prodigious jumping, which enabled him to threaten the sun and to overleap the ocean, and his strength, capable of uprooting a mountain. Despite these points of superiority, he chose to be the servant of humans.

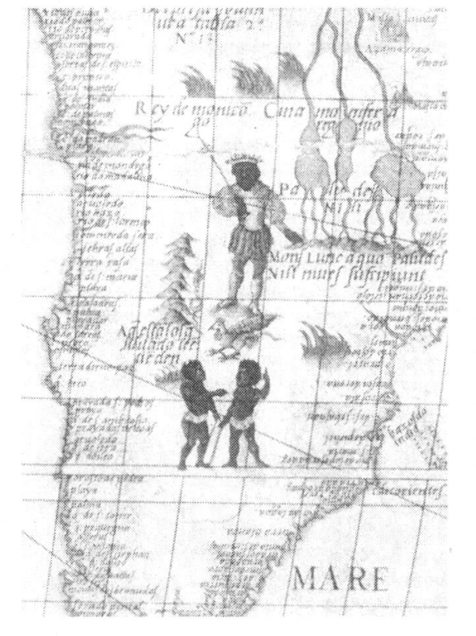

13. Sebastian Cabot's World Map of 1544 includes a representation of pygmies in the African interior, engaged in leisurely conversation, with classical gestures, and equipped with sticks – evidence of their tool-using status (shared, at the time, in many accounts and images of apes). Above them the King of Kongo is attired – accurately for the time – in European style (see p. 67).

14. At the time of Saartje Baartmann's London visit in 1810, a 'broad-bottomed coalition' under Spencer Perceval ruled in Westminster, shunned by the leading politicians of the time. The opportunity for satire was irresistible at the arrival of 'a new Addition to the Broad-Bottom Family'. (left)

15. The Dutch anatomist Petrus Camper (1722-89) became interested in craniology as a result of his passion for accuracy in anatomical drawing. He believed that 'facial angle' – the extent of protrusion of the jaw – demonstrated 'a striking resemblance between the race of Monkies and of Blacks'. (below)

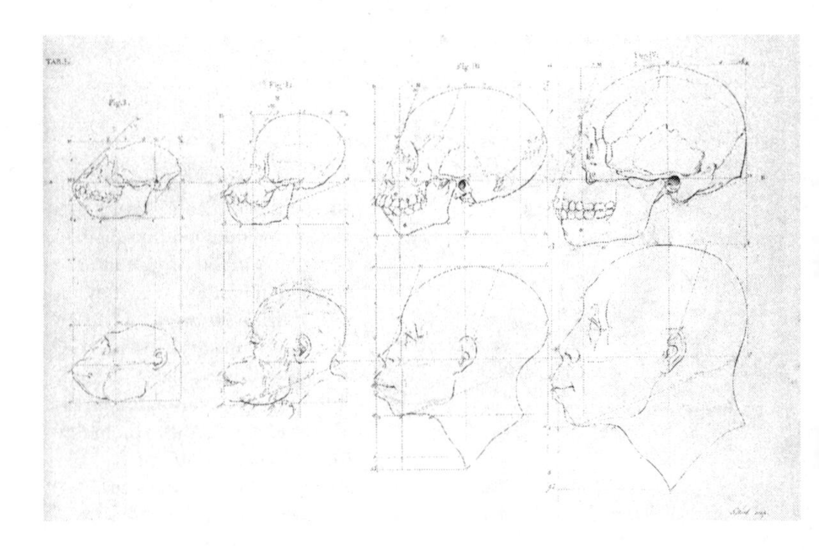

making one of his more fruitful suggestions: that the proliferation of monsters in the legends of the world is the result of the difficulty group members have in describing outsiders. Monstrosity, according to Gobineau, is alterity misrepresented. One of the reasons why our modern conception of humankind is so unusual is that normally, in history, people define their own group as human and ascribe varying degrees of beastliness to those outside it. Strangers, even if they are indistinguishable in appearance from group members, are credited with physical oddities, which are then exaggerated and distorted into shapes of otherness, or are supposed to appear human by disguise.

The monsters originated, perhaps, as creatures of credulity, fabrications of fear. But reason intervened to reclass them as a kind of normality. From the fifth century to the thirteenth, the way Western intellectuals thought about them was dominated—even determined—by the prestige of St Augustine. He was sceptical of the existence of all the monstrous progeny of man which peopled legend, but experience taught that such miscreations or freaks of nature did exist: they should be treated like other human deformities—those with hunch backs, extra toes, dwarfish stature, and so on—as beautiful to God. You could—to mention a few of the peculiarities ascribed by ancient and medieval biology to 'monstrous races'—have a dog's head or a tail, or a mixture of male and female anatomical attributes, and still be human. Reason, according to Augustine, was the only criterion of humankind, 'since if we did not know that apes, monkeys and sphinxes were beasts rather than men' we 'could easily misrepresent them as human' on grounds of their physical resemblances.

The next revolution in thought on these questions was initiated in the thirteenth century by Albertus Magnus—the teacher of St Thomas Aquinas and a towering figure of the age: one of the great moulders of the high-medieval intellect. In a radical departure from Augustine's thinking, he insisted that reason preferred to dwell in perfect bodies, and physical monstrosity was a mark of the beast. Striking physical peculiarities, formerly acceptable in groups classifiable as human, now suggested a non-human or sub-human nature, an irremediably bestial status. This was an unwarranted presumption but it presumably arose as a supposedly scientific inference from the fact that idiocy and deformity sometimes coincide; sometimes, moreover, they are driven into alliance by social ostracization of people who are physically odd or mentally eccentric. It is unsurprising that these facts should inspire the theory that rationality is a property of normality.

How far could the theory be made to extend? Was black or brown skin pigmentation an incapacitating deformity? What about woolly hair or thick lips? Or the hairless visages and lank hair of most native peoples of the New World? Or the stature of pygmies or the protuberant buttocks of female 'Hottentots'? What about flat noses? These were particularly problematic since according to the masterfully authoritative *Etymologies of St Isidore*, 'simia'—the commonest word for apes and monkeys—meant 'flat-nosed'. Over people who evinced any or all of these features, questions concerning their admission to human status hovered when Europeans first encountered them. The language of beastliness sprang easily to the pens of writers who described them. From the fifteenth century onwards, these problems were tested in Western minds by explorations in sub-Saharan Africa and deepening encounters with people who were black.

Colour-coding the Human Body

Some prejudice against blacks had, perhaps, accumulated in Europe before the fifteenth century. Black was medieval artists' colour-code of preference for demons: when the Portuguese chronicler Gomes Eanes de Zurara first saw black slaves in 1440, he at once compared them with 'denizens of the infernal regions'. Thick lips, moreover, according to medieval pop-psychology, signified a lustful disposition: and lust was a demonic device for subverting reason—it was, in part, what made wild men wild and beasts beastly. To judge from a line in the fourteenth-century Aragonese compilation, the *Libro del conoscimiento de todo el mundo*, or 'Book of Knowledge of All the World', black skin colour was suspected in some circles as a disqualification for fully human status: 'although the Nubians are negroes,' says the author, with apparent surprise, 'yet they are endowed with reason, like us.' Maghribi contempt for blacks may have communicated itself to Europeans: it certainly infected some blacks. The ruler of Mali at the time of Ibn Battuta's visit to that realm in the 1350s was black, but perceived himself as white, evinced disdain for his subjects on grounds of their colour and characterized his kingdom as 'a white spot on a black cow-hide'—meaning that it was an oasis of civilization surrounded by savagery.

Nevertheless, the balance of evidence tends to support the view that in medieval Europe blacks were generally regarded as the equals of white people, without any of the distortions of later racism. Christendom in that period, until well into the fifteenth century, knew blacks only in small numbers, usually as individual slaves, purchased from North African traders. Slaves were of all colours and there was no particular association between blackness and inferior social status. There were no slave plantations: most slaves were domestic servants. The way their masters thought about them was influenced partly by the emotions which arise from personal contact, including, in many cases, affection and esteem, and partly by the heritage of classical antiquity, when slaves discharged reputable duties as, for example, teachers and secretaries. In the Greek comedies of Menander and the Roman comedies of Terence and Plautus, they were commonly depicted as the intellectual superiors of their masters, whom they manipulated, hoodwinked, and exploited, without malice and in a spirit of fun. The normal relationship was more like that of Jeeves and Wooster than that of Jonathan Corncob and his 'hen negro'. One of the comedies of Gil Vicente, the most successful Portuguese playwright of the Renaissance, depicted a haplessly, hopelessly lovesick black slave who wanted to be white in order to attract the goddess Venus, but most of his black characters are in the classical slave-mould: cheeky, sometimes rascally, but ultimately cheery and reliable.

Enslaved blacks might occupy responsible positions in the households they served, or work in specialized, skilled occupations. In fifteenth-century Portugal possession of a band of black musicians was a sign of social prestige. A black singer performed at the wedding of the king's daughter in 1455. Because there were black slaves there were, necessarily, growing numbers of black freedmen, who seem to have had access to every level of society. In societies more sensitive to distinctions of rank than of race, blacks could be recognized as noble. In 1475, Ferdinand and Isabella appointed a leader of the confraternity of free blacks of Valencia on the grounds that 'you are of noble birth among the said blacks'.

Most images of blacks, moreover, were positive. Most Europeans probably never saw a real, live black person. They would, however, see images of black kings in depictions of the Adoration of the Magi, one of whom was conventionally depicted as black. Negritude, therefore, carried associations of regality, wisdom, and the privilege of one of the earliest of divine revelations about the nature of Christ. The story, told in the Acts of the Apostles, of St Philip's conversion of the Ethiopian eunuch, who then, presumably, introduced Christianity to his native land, reinforced these associations. It was, of course, a myth: Christianity became the court religion of Ethiopia in the fourth century, shortly after the conversion of the Roman emperor Constantine. Nevertheless, the fact that a Christian community of great antiquity existed in a black realm in the depths of Africa was never quite forgotten in the West, and in the fourteenth and fifteenth centuries visiting clergy from Ethiopia occasionally appeared in Rome. Sixtus V, indeed, established the Church of San Stefano dei Mori as a haven and house of study for them.

By the late Middle Ages, Europeans therefore had high expectations of the level of civilization and kindred-feeling they might find if ever they established direct links with black Africa. These anticipations were heightened, meanwhile, by the influence of a further myth: that of 'Prester John', a Christian monarch of immense wealth and power, whose dominions lay beyond the territories of Islam and who—if he could be contacted—would join the Crusades. In the first known occurrence of this legend, in the work of Bishop Otto of Freising at the time of the Second Crusade, Prester John was located 'in Nubia', raising the presumption that he was black. Medieval geography was generally vague, and later versions of the story located the Prester in India (where there were Christian communities but no Christian potentate) or central Asia (where some Mongol chiefs had adopted Nestorian Christianity by the late twelfth century).

Nevetheless, with the swings of fashion and the growing contacts with Ethiopia, the association of the fabled monarch with the real Christian empire in the Horn of Africa grew stronger again in the fifteenth century until, by the early sixteenth century, Europe and Ethiopia became directly linked by the development of the sea-route around the Cape of Good Hope. The identification of Prester John with the ruler of Ethiopia became taken for granted. The first account of a European embassy to the country, published in 1520, was actually entitled, *The Prester John of the Indies*.

Late medieval maps literally gleam with Europeans' high expectations of the black world and of the civilized habits of its peoples. Black Africa appears dotted with gilded cities and richly arrayed monarchs. Indeed, the gilding lavished on such images was made from gold-dust imported from black Africa via the Sahara. Europe's main source of gold was the Bure region of upper Sengambia, from where it was traded to the Mande merchants of the immensely rich empire of Mali, then exchanged for Saharan salt with Maghribi caravaners who carried it north across the desert to Mediterranean ports. The rulers of Mali kept the nuggets as tribute: in consequence, at the height of their power in the fourteenth century, they were able to maintain a lavish court, a Muslim clerical and literary establishment 'imported' from the Maghrib, and a large army. Knowledge of Mali burst upon the world in the 1320s, when King Musa made a pilgrimage to Mecca with a caravan reputedly of 800 camels, laden with gold. His gifts to the shrines he visited on the way were so lavish that he caused inflation in Egypt, variously estimated at between 10 and 50 per cent. His effulgent image invaded European imaginations. In fourteenth- and early fifteenth-century maps by Christian cartographers, you can see rulers of Mali, dignifiedly bearded, crowned, and enthroned with orb and sceptre, or brandishing gold nuggets, in depictions of majesty in European style. In some ways, this white willingness to concede equality of status to blacks survived the era of the inception of direct European contact with black Africa in the fifteenth century. When Portuguese adventurers became involved in the affairs of the kingdom of Kongo in the 1480s, the monarchs of Portugal showed the same courtesy to their Congolese 'cousins' as to fellow monarchs in Europe. A Congolese prince became an archbishop.

Yet familiarity bred contempt and the more Europeans knew about black cultures, on the whole, the less they liked them. Ironically, the prejudices white people conceived of blacks were, in part at least, the outcome of excessively sanguine expectations. The first black societies known in their own environments to European explorers were those of the Imraguen and Znaga of the western Sahara, encountered by Portuguese expeditions in the 1430s. These peoples had, by European standards, some of the most rudimentary technology, and the most modest material culture imaginable. They were fisher-folk—the marginalized remnants of a pre-Berber population, whom successive migrations and invasions from the east had left languishing on the desert's edge, clinging to the rim of the ocean. The Wolofs of Senegambia, with whom Europeans became acquainted in the 1450s, were much richer and more sophisticated, but still disappointing to minds full of stories of Prester John and Musa of Mali. The first white ethnographer of the region, the well-born Venetian Alvise da Mosto, who left an intriguing account of a voyage along the Gambia as a passenger on a Portuguese vessel, was remarkably objective and non-judgemental, but it is evident that he found the people comically primitive. The most impressive chief he met was called Budomel. He could barely ride a horse and asked the Venetian for advice on 'how to satisfy more women'. Da Mosto also reached an upriver-outpost of the empire of Mali. It seemed a far cry from the gilded realm imagined by the mapmakers: a land of poverty and ignorance, mud huts rather than adobe palaces. Indeed, to the great misfortune of future black–white relations, Mali was then in decline. The rise of the rival empire of Songhay to the south-east had engaged it in war, diverted its trade, impoverished its revenues, and captured one of its principal cities—Gao, on the lower Niger. The image of the ruler of Mali changed in the fifteenth century. Instead of the stately monarch of the cartography of the previous century, he became a figure of fun—a stage nigger, dangling a grotesque phallus. Black slaves in the drama of the time were not depicted in terms familiar from the plays of Plautus as clever rogues, but became re-cast as comic lowlife.

As slavery came to be practised on a large scale, and to be associated exclusively with blacks, it exacerbated prejudice. The effect probably happened many centuries earlier, in Islam, where the happenstance that so many slaves were black may have implanted the idea that blackness was a feature of the inferior. Black poets resented it, but were unable to reverse it. Suhaym, who died in 660, protested that though his skin was black, his soul was white. A couple of generations later, Nusayb ibn Rabah recommended himself as 'a keen-minded, clear-spoken black' and therefore 'better than a mute white'. Ibn Khaldun explained blacks' suitability for slavery on the grounds that their attributes were 'quite similar to those of dumb animals'.

Relegated to an inferior category of humankind, blacks, perhaps, became a precedent for the re-classification of those other creatures with whom observers sensed some kindred. Some of these we now know as human; others we class as apes. Since baboons, chimpanzees, gorillas, and orang-utans all qualified for consideration, at least, of their suitability for inclusion in the human family, the genuine puzzlement caused by encounters with the unfamiliar is plain enough; perhaps therefore we should not judge too harshly the uncertainties of European observers and commentators, in the same period, over whether pygmies or the people called 'Hottentots' and 'blackfellows', for example, might, on first acquaintance, belong to some non-human species.

'Hottentots,' according to Sir Thomas Herbert in 1634, 'have no better predecessors than monkeys'. They were characterized by other visitors as 'beasts in the skin of a man' or 'filthy animals who hardly deserve the name of rational creatures'. The literature is confusing, since much of it evidently relates not to the Khoi, who correspond to the people now usually identified as 'Hottentots', but to the San or Bushmen. The former are a pastoral and commercial people who made, on the whole, a favourable impression on early Dutch colonists to the Cape. The latter are hunter-gatherers, who preceded the Bantu and Khoi in southern Africa and were treated as enemies by them. The physical peculiarities of San women—bulging buttocks and amply developed nymphae and labia which formed an 'apron' over the vulva—exemplified, for many observers, the human insufficiency of 'Hottentots' and constituted evidence for relegating them to a sub-human category. In 1810, the 'Hottentot Venus', Saartje Baartmann, was exhibited in London and Paris and her steatopygous posterior and hypertrophic 'apron' could be seen by all. Her sexual organs, preserved after dissection at her death, remain in the Musée de l'Homme of Paris.

Meanwhile, if any group were capable of exceeding the San in their ability to attract learned contempt, it was Australian aborigines. In the opinion of William Dampier, author of the first description of some of them, published in 1697, the Hottentots 'though a nasty people' were 'gentlemen' by comparison. His greatest revulsion was from the poverty of their material culture but 'setting aside their human shape' he also found them physically repellent, 'having no one graceful feature in their faces'. The aborigines attracted little scientific study in the eighteenth and early nineteenth centuries, but their relegation to inferior status seemed to require no endorsement from science: it struck those who met them as obvious. Despite the British government's pious injunctions to subjects 'to live in amity and kindness' with them, ruthless persecution and exploitation ensued. In contrast to the practices adopted among Native Americans or Maoris, British imperialists never bothered to make treaties with Australian aborigines. Colonists hunted such people without qualms. Early painters of the 'black-fellows' were divided in their perceptions: there are some noble-looking savages among their subjects; more commonly, however, the aborigines are simian creatures, crawling on the earth or capering in the trees.

In the age of encounters, doubts about who was and who was not human evidently arose from more than skin colour. To a great extent, the connotations were cultural: the subject of the next chapter. As far as physical criteria were concerned, pigmentation was only one trigger of doubt. On its own, it was an unreliable indicator, since there were so many gradations between black and white that it was impossible to draw the line. The San are actually rather a light brown in colour; so are the West African Fulbe. The Spanish and Portuguese languages in the eighteenth century—spoken in the most racially diverse empires of the day—developed dozens of different terms to designate various degrees of colour. Most of the people these terms described were themselves examples of varying patterns of mulatto-hood or mestizaje—engendered by differently pigmented parents. The fact that such unions were fertile and had no discernible effects on the fertility of their offspring demonstrated the value of a category capacious enough to embrace all those who could fruitfully embrace each other. Since, however, the question of whether humans and apes could successfully interbreed was still not satisfactorily resolved in the eighteenth century, the problem of how narrowly to circumscribe humankind remained open, even by the standards of this apparently minimal criterion. The search was on for a scheme of classification which set skin colour in the context of an objective, quantifiable set of criteria which would settle doubt about who was, and who was not, part of humankind. The long history of black–white interactions had by now dispersed whatever positive inclinations whites had once had towards blacks, and the latter would have to struggle—and enlist liberal-minded whites on their behalf—to vindicate their human status in the new, Western-dominated science of the modern era.

Room for the Untermenschen: Among Races, Species, and Sub-species

In nineteenth-century and early twentieth-century Europe, classification of humankind into races was thought to be scientific, by analogy with botanical taxonomy. William Lawrence, whose influential lectures on anatomy were delivered in London in 1817, revived the claim first made by Albertus Magnus: 'physical frame and moral and intellectual qualities', as he put it, were mutually dependent. 'The distinction of colour between the white and black races is not more striking than the pre-eminence of the former in moral feelings and in mental endowments.' Various methods were proposed—according to pigmentation, hair type, the shape of noses, blood types (once the development of serology made this possible), and, above all, cranial measurements. This last method was devised by the late eighteenth-century Leiden anatomist, Pieter Camper, who arranged his collection of skulls 'in regular succession', with 'apes, orangs and negroes' at one end and central Asians and Europeans at the other. There was obviously an underlying agenda: a desire not only to classify races but also to justify disparities of power by ranking them in terms of superiority and inferiority. Hence the emphasis on the shape and dimensions of the skull, which were alleged to affect brainpower. (There is no real connection: Turgenev had a large brain, but Anatole France had one of the smallest ever measured. Men and women are, on average, equally clever, despite general differences in brain size. Human brains have been getting smaller for about the last 15,000 years—for unknown reasons, perhaps connected with our abandonment of the mentally demanding complexities of foraging ways of life; and though we have made no discernible progress in intelligence over that period, we seem to have remained on average, as good or bad as ever at thinking.)

It was one thing to assert the disunity of humankind, another to devise a theory which made it credible. The most obvious option was the theory of polygenesis, according to which creatures loosely classed as human had emerged separately, whether by nature's laws or heaven's command. This had first been advocated in a work published in 1655 by the Calvinist theologian Isaac de la Peyrère, as a solution to the problems of the diversity of humankind and, in particular, of the origins of the peoples of the New World. Were they the lost tribes of Israel? Had Noah settled in Brazil, as one early seventeenth-century authority argued? Or had the first settlers come from Asia, according to the theory in which the Spanish Jesuit José de Acosta pre-empted the discoveries of modern anthropology? At the time, all these hypotheses seemed equally improbable. La Peyrère suggested that the universal paternity of Adam should be understood metaphorically, making credible the origins-myths that so many Native American peoples cherish: that they were sprung 'from their own earth'. The theory was dismissed by no fewer than twelve respondents in its year of publication. It was as contrary to the religious orthodoxy of its day as it was to the Darwinian orthodoxy of a later age. Its periodic revivals were, on the whole, feeble and of limited appeal.

Degeneracy was another potential theoretical framework for understanding supposed racial inferiority. The popularity of the term among nineteenth-century anthropologists is intelligible in the context of a 'discourse' of degeneracy, employed to explain all sorts of exceptions to progress: criminality, psychiatric pathology, economic dislocations, national decline, and, ultimately, the supposed 'degeneracies' of 'modern art'. In the late nineteenth century, says its chronicler, Daniel Pick, degeneracy 'slides over from a description of disease or degradation as such, to become a kind of self-reproducing pathological process—a causal agent in the blood, the body and the race—which engendered a cycle of historical and social decline perhaps finally beyond social determination.' In 1870, Henry Maudslay, professor of medical jurisprudence at University College, London, united some of the themes:

> When the development of the 'brute brain' within man, he reasoned, 'remains at or below the level of an orang's brain, it may be presumed that it will manifest its most primitive functions. . . . We may without much difficulty, trace savagery in civilization, as we can trace animalism in savagery; and in the degeneration of insanity, in the *unkinding*, so to say, of human kind.

Among supposedly degenerate groups of humans, the concept of 'gradation' offered an apparent means of measuring degeneracy. The term was coined by Charles White in the 1790s, when he produced an index of 'brutal inferiority to man' which placed monkeys only a little below blacks, and especially the group he called 'Hottentots', whom he ranked 'lowest' among those who were admissably human. More generally, he found that 'in whatever respect the African differs from the European, the particularity brings him nearer to the ape'.

Popular in the eighteenth and early nineteenth centuries was the theory now associated with the name of Jean Baptiste de Lamarck. As he formulated it in 1809, biota adapted to their environments and such adapted characteristics were passed on by heredity. As an explanation of 'racial' differences, this had long been a theme of popular science. Boswell recorded a conversation in which Samuel Johnson explained why blacks are black: ever-deepening sun-tans had been transmitted to their progeny over many generations. Darwin—whose theory of evolution is now recognized to be incompatible with Lamarck's—actually endorsed his predecessor's views. In deference to Lamarck, Darwin advised young women to acquire 'manly skills' before starting families. (Nonetheless, one of the advantages of his own account of evolution was that it did not rely on the dubious claim that acquired characteristics are heritable.) The Lamarckian idea has never quite vanished from the repertoire of scientific explanation, though the arguments of Darwinism have tended to eclipse it. Experimental data do not seem to support it and common observation is against it. You may sit in the sun all your life, but your children will be no darker for it.

None of these theories—of polygenesis, of degeneration, of climatic determination of heritable characteristics—could ultimately be reconciled with the Darwinian orthodoxy which gradually became dominant. Polygenesis was excluded by Darwin's basic assumption—that all life had a common origin; degeneration was incompatible with the implicit progressivism of the doctrine of the 'survival of the fittest'; the heritability of acquired characteristics (though Darwin personally did not wish altogether to rule it out) was an unnecessary hypothesis in the light of natural selection. But it was possible to reconcile evolution with other varieties of what we might call species-theory: the doctrine that there were more than one species in the genus *Homo*, was, after all, implicit in Darwin's own account of the descent of man, which postulated earlier hominid types. As discoveries of fossil hominids accumulated, and palaeoanthropology assigned them classifications in the genera of 'Homo' or 'Pithecanthropos'—man or ape-man—*au choix*, it seemed perfectly reasonable to suppose that, just as there had been more than one species of *Homo* in the past, so there might be still: a 'missing link' might yet live and await discovery.

Indeed, such para-human types might already be known, awaiting nothing more than appropriate classification. Edward Long had justified slavery in 1774 on the grounds that blacks were differentiated from other peoples —inter alia, by a 'narrow intellect' and 'bestial smell'—so as almost to constitute a different species. Henry Home in the same year went further: humans constituted a genus in which there were numerous different species, of which blacks were an obvious example. At first, this view was generally rejected for the reason we have already encountered: it was obvious that humans of all kinds are capable of breeding with one another; but the compulsion to find a way of characterizing the diversity of humankind consistently with the prejudices of the times was keenly felt among scientists. Evolution opened up new possibilities.

To Darwin, races were 'sub-species' or potential species: blacks and whites, for example, might eventually become separate species, if kept apart from one another, by analogy with the separation of different species of gibbon, say, or tern, or of closely related felines. To many other scientists of his day, 'human' was a misnomer for races already divided from the human norm by unbridgeable chasms, if they were not actually products of polygenesis—the 'separate creations' which Darwin denied. According to Samuel Morton of Philadelphia, who died while Darwin was at work on *The Origin of Species*, Native Americans were unrelated to people in the Old World: they had evolved separately in their own hemisphere. The findings *à parti pris* of Josiah Nott and George Gliddon—that blacks were more like gorillas than full-ranking human beings—appeared a year before Darwin's work was published. In the 1860s, John Hunt, founder of the British Anthropological Society, endorsed the similarity between blacks and apes and attributed cases of high attainment among blacks to exceptional instances of interfertility among separate species—admixtures of white blood (which, he thought, were nonetheless non-viable in the long run). Meanwhile, his sometime associate, John Crawfurd, revived the notion of polygenesis, while explicitly denouncing the view that distinct human species could be ranked on grounds of colour.

The Comte de Gobineau died in the same year as

Darwin. Relying more on what was then beginning to be called anthropology rather than on biology, he worked out a ranking of races in which 'Aryans' came out on top and blacks at the bottom. 'All is race', concluded a character in one of Disraeli's novels. 'There is no other truth.' Gregor Mendel, the kind and gentle Austrian monk whose experiments with peas established the foundation of the science of genetics, died two years later. The implications of his work were not followed up until the end of the century, but, when drawn, they were abused. With the contributions of Darwin and Gobineau, they helped to complete a supposedly scientific justification of racism. Genetics provided an explanation of how one man could, inherently and necessarily, be inferior to another by virtue of race alone. To the claim that this represented a new departure in the history of human self-perceptions, it might be objected that racism is timeless and universal. What the nineteenth century called 'race' had been covered earlier by terms like 'lineage' and 'purity of blood'. No earlier idea of this kind, however, had the persuasive might of scientific racism; nor the power to cause so much oppression and so many deaths.

In partial consequence, the first half of the twentieth century was an age of empires at ease with themselves, where critics of imperialism could be made to seem sentimental and unscientific, in a world sliced by the sword and stacked in order of race. Just when white power was at its most penetrative and most pervasive, scientific theory helped to ram it home. Inferior races were doomed to extinction by natural selection; or could be actively exterminated in the interests of progress.

It was, I suspect, precisely because racism was so strongly and widely upheld, that an inclusive view of humankind triumphed in the same period. Racism provided ample justification for the victimization, persecution, oppression, and extermination of some groups by others. It became unnecessary—even for advocates of Nazism or apartheid—to insist that different human groups constituted different species, sub-species, or potential species: they could be fully human without redemption from the taint of inferiority. More recently, the defeat of the 'master-race' and the retreat of white empires have relieved us of the need for racist explanations of the way the world is.

Meanwhile, biology has made racism indefensible. Not only are there no inferior races: there are no races; there is practically no racial differentiation among humans. Although we may look different from one another, the genetic difference between the most widely separated humans is tiny, by comparison with other species. The same science has exploded the notion of human 'subspecies'.

We have, in short, learned to be inclusive the hard way: by killing each other. In a sense, the Untermenschen have been harried out of existence. Inclusiveness has not been scientifically discovered—it has been painfully imposed. Today we need it more than ever. We need an inclusive self-perception in order to empower us for peace in multi-cultural societies and a multi-civilizational world.

CHAPTER 3

HUMAN BEING OR BEING HUMAN?

The Quest for a Cultural Solution

Potentially inclusive definitions of humanity are traceable, as we have seen, in traditions which began in the first millennium BC in Indian, Greek, and Chinese texts; but all these civilizations—and others with similar concepts—admitted the existence of deficient or humanly imperfect categories within humankind, including those of women and 'barbarians'. Moreover, they assumed the existence of sub-human species in the interstices of the heirarchy of nature, between those that are fully human and those that are utterly non-human. The problem of where particular beings or groups fitted into this scheme of classification was therefore unresolved. The story of the last chapter shows how hard it was to resolve it by looking at people's physical characteristics—whether 'normalcy' of bodily proportions, or cranial dimensions, or skin colour, or type of body hair.

An alternative approach lies through the study of culture. Perhaps human is as human does. By setting thresholds of behaviour, the fully human could be distinguished from the sub-human, near-human, and utterly non-human. In this chapter we trace the history of efforts to identify human credentials in culture: ways of behaving and relating that—according to self-appointed arbiters—admit those who display them to the human community.

The consequences can be followed in the history of broadening encounters between cultures, as peoples of a previously unanticipated diversity confronted one another for the first time. This part of the story occurred mainly in the period of earth-girdling navigation, which began in Western Europe about five hundred years ago and which provoked challenges to just about everybody's notion of the nature and limits of humankind. This chapter, therefore, focuses on a relatively late and brief period and on predominantly Western experience—but with glances back in time and across the world for comparative purposes.

The Sociology of Savagery

When Gulliver was shipwrecked on Houyhnhnm Land, he hoped to be able to throw himself on the mercy of some savages, but encountered instead 'strange creatures', who walked on their hind legs and 'held their food between the claws of their fore feet'. Their skin was 'buff coloured'. They had little hair, except on their heads, where it was 'of several colours, brown, red, black and yellow', and in clusters round their pudenda. 'I never beheld in all my travels so disagreeable an animal, or one against which I naturally conceived so strong an antipathy.' These brutes—whom the reader would recognize as men, wryly described in terms drawn from travellers' accounts of apes—were employed as beasts of burden by the noble, learned horses who ruled the country. The choice of horses as heroes was determined by a circumstance obvious to Swift's readers in his day. In standard Renaissance works of moral philosophy, horses were commonly cited as examples of 'irrational creatures' to contrast with humans, defined as rational. Gulliver became the servant of a horse, whom he astonished with tales of the topsy-turvy world he came from, where humans were 'the only governing, rational beings' and horses were treated as beasts.

The body shapes of the two species of Houyhnhnm Land were no guide to their rationality. To judge how well they could reason, the traveller had to observe their social lives—their manners, relationships, laws, customs, common pursuits, and collective achievements. The horses, who considered themselves 'the perfection of nature' were courteous and collaborative and possessed the means of communicating by language (which the humans had too, but the horses did not recognize their utterances as true speech). They practised no vices, in which they could see 'no use or necessity', fought no wars, for which they could perceive no rational cause or motive, and did nothing for which good reason could not be shown. 'Their grand maxim' was 'to cultivate reason and to be wholly governed by it.' The humans, on the other hand, behaved abominably, but a wise horse could no more blame them than he could blame 'a sharp stone for cutting his hoof. But when a creature pretending to reason could be culpable of such enormities, he dreaded lest the corruption of that faculty might be worse than brutality itself.'

Swift was asserting, in effect, that human is as human does. In the West Swift's claim was relatively new in its day. Ancient China, however, anticipated modern controversy about whether human status is fixed in us by nature or attainable by action—in particular, by behaving socially. Mencius's maxim suggests this: 'to lack a father and lack a ruler' is to be a bird or beast. So does Mo Ti's account of the origins of human culture: 'birds beasts and insects,' he observed, 'use feathers and hair for clothing, hoofs and claws for sandals and shoes, water and grass for food and drink' and so do not need to organize socially for survival, whereas human deficiency in these respects drove 'man to till and woman to spin'. In the third century BC, Hs'un Tzu thought a social gift was the essence of humankind: animals had perception but no sense of justice; their groups were not therefore communities in the fully human sense. Men could exploit stronger creatures because they were able to form societies and act collaboratively. A thousand years later Tai Ch'ih maintained a similar doctrine: social tendencies are peculiar to humankind; our antisocial tendencies are shared with other animals. Indeed, it seems to have been widely supposed in ancient China that culture was the defining characteristic of humans, because only humans had it. Only during the Tang dynasty did this assumption begin to be questioned in surviving texts, noting similarities between human society and those of bees and ants. Kuan Yin in the eighth century assumed that humans had learned social order from bees and war from ants. According to the tenth-century *Hua Shu*, the first human civilization resembled ant society, with its political unity, organic self-perception, and food storage and regulation.

These formulae immediately reveal the difficulty of a cultural approach to the problem of definition: many non-human creatures—from ants and bees to dolphins and whales—are more gregarious and live in more tightly knit societies than ours. Others, such as dogs and other domesticates suitable as pets, become more fully part of human society than many humans: there are no drop-out dachshunds or loner lurchers. Nevertheless, there are forms of relationships and kinds of culture which are peculiar to particular species. So it might be useful to discard or set aside the problem of defining the nature of 'human being', and turn instead to that of 'being human': of defining humankind in terms of human culture, rather than human nature.

In the Western tradition enquiry along these lines began long before any theory to the effect that human nature could be identified culturally, at the beginning of the fifth century AD, when St Augustine dismissed physical criteria as the defining characteristics of humankind. At first, the search for what makes humans human remained focused on Aristotle's essentially mental characterization of people as rational animals. For most of Augustine's successors, the best practical way of determining whether creatures have reason was to examine what was known or alleged about them for outward signs of supposedly rational behaviour, such as clothing or laws, technical skills or artistry. So debate got dragged away from the human body and the human mind to examination of human society. The most basic criterion for admission to the ranks of humankind was the practice of what observers could recognize as social life.

On these grounds, throughout the Middle Ages, the 'cynocephali'—the dog-headed men whom some authorities identified as baboons—were admitted by most writers, who referred to them as candidates, at least, for admission into the ranks of humankind: the good impression baboon 'government' made on Richard Jobson was part of this tradition. Nor were the known categories fixed. Humankind was a class you could slip in or out of: there were degrees of social development—or, as we should now say, civilization—and people who occupied the lower ranks shared features with the beasts.

This comparative ethnology seems to have been implicit in the way ethnographers in Latin Christendom referred to neighbouring peoples unprivileged in these ways; its first full and explicit formulation, however, occurred only in the late twelfth century, in the work of Gerald of Wales, a

Normanized scholar engaged in a surprisingly 'modern' quest for his Celtic roots. Convinced that secularism was civility and transhumance was savagery, he was also susceptible to the myth of Arcady: he liked to picture his fellow Celts as revelling in a bucolic idyll. He depicted the whole of Wales—which, in his day, had substantial farmed and urban areas—as pastoral, but with sidelines in banditry and rapine. He condemned the Welsh as incestuous and promiscuous, but with the conventional virtues of a shepherd-race among whom 'no one is a beggar, for everyone's household is common to all. They prize liberality, especially generosity, above all virtues.' For deeper enlightenment about the Celtic past, he turned to Ireland. Its people, he found, were hairy infidels, 'a wild race of the woods . . . getting their living from animals alone and living like animals', astonished by the sight of bread. Balancing the evidence of bestiality and humanity, Gerald developed a theory of social development. The Irish, he concluded, 'have not abandoned the first mode of living—the pastoral life. For when the order of mankind progressed from the woods to the fields and from the fields to the towns,' the Irish preferred 'the life of the woods and pastures' to the labours, treasures, ambitions, rights, and responsibilities of civilization.

It was an option not unlike that chosen by Jobson's baboons. Later characterizations of Europe's Celtic fringe reflected the same prejudices—the same conquistador-values. Because they led a pastoral life, wore pelts, and built no cities, the Irish in the sixteenth century were easily dismissed as savages by their English would-be conquerors. Their Spanish allies thought little better of them. Francisco Cuellar, who survived a shipwreck of the Armada of 1588 and left an account of his escape across Ireland, casually referred to his hosts as savages; despite their Catholic avowals, he could hardly recognize them as co-religionists. The Highland and Island Scots in the same period faced a similar fate: while the English conquered the Irish, the Lowlanders of Scotland were engaged in a similar campaign against their neighbours to the north and west. James VI—soon to be James I of England—abhorred those 'most savage parts' of his kingdom and offered lavish fiscal concessions to conquistadores willing to reduce them 'to civility'. These were not, however, prejudices forged by racial stereotyping. Any remote, rural community might attract the same sort of metropolitan contempt. Henry VIII thought something similar about the poor of Lincolnshire—'the most brute and beastly of the realm'—when they had the audacity to rebel against his tyranny.

The explorations reported by Gerald of Wales were part of a great enterprise, launched in his day and lasting for centuries. Ethnographers in Latin Christendom strove to comprehend the peoples of their own internal frontiers, the folk of forest, bog, tundra, and mountain, the inhabitants of the under-studied, under-evangelized recesses and edges of Europe. The results of the enquiry were equivocal: accounts of Europe's internal 'savages' combined exemplary and cautionary tales. A contemporary of Gerald's—writer of a pilgrim guide to Compostela, who called himself 'Aiméry Picaud'—was fascinated by the mountain ways of the Basques: buggery with mules and river-poisoning. The Cistercian Gunther of Pairis, who put the history of his times into verse, found the peoples of the Pripet marshlands 'crude-mannered'. Yet there were ways of assuaging the savagery of one's neighbours. Those who lacked cities or espoused a pastoral or foraging way of life were meaner, in the heirarchy of societies, than the medieval 'first world' of farms and towns; but their humanity was unquestioned. Indeed, their simplicity could be seen as virtuous: in terms of the classical heritage, it recalled the 'Golden Age' of silvan innocence of which Greek and Roman poets sang, which preceded the fall of Saturn. In Christian terms, it suggested the innocence of Eden.

Medieval moralists therefore often extolled the 'good barbarians' whose values, uncorrupted by ease and wealth, were examples which could be exploited to challenge or chastise civilized vices. It was particularly useful in the case of pagan peoples, who could, in selective cases, be said to behave better than Christians despite their lack of the light of the gospel. The model of the Good Samaritan was irresistible to some writers who wished to echo, for their own times, Christ's criticism of the society that surrounded him and his recognition of the outsider's potential for virtue. The eleventh-century German historian Adam of Bremen praised the Prussians—a now extinct pagan, Slavic, pastoral community beyond the eastern edge of the Christendom of his day—because 'unlike us, they despise gold and silver as dung'. Marco Polo invented an 'innocent Tatar': a particularly effective example because particularly shocking; for the Tatars' reputation for brute ferocity was a commonplace—a bogey-image—of his time. A good deal of the imagery and attributes of the virtuous pagan, developed in this period, later became part of the identikit of the 'noble savage'.

Up to this point in the story, examples of the sorts generated by European experience could be multiplied from within other major civilizations of Eurasia. China, for instance, had its own internal barbarians—the Li, the Miao, the Nosu, the Hakka, the Peng-min, and many smaller or more marginal groups—who could be treated in literature with the same mixture of condescension, repugnance, and appropriation for didactic purposes; sometimes they were depicted as 'packs of beasts', expected to 'grasp and bite' or as demon-like creatures of implacable savagery; otherwise they could seem equally convincing as exemplars of natural virtues, practising Confucian austerity without benefit of instruction; or in the case of the Hakka—who were a Han people, resembling metropolitan Chinese closely in ethnic origin and culture—they might be self-represented as models of loyalty to the empire. According to their greatest apologist, the late eighteenth-century official Hsu Hsu-Hseng, the Hakka were 'diligent, thrifty, courteous, modest, elegant and polished'—throwbacks, in short, to a Chinese golden age of virtues long since corrupted in the heartlands of China. Chinese expansion, however, remained largely confined to areas contiguous to China, and ethnographic literature never acquired the breadth of reference which later became available to Europeans. It was a Confucian principle, moreover, to attract barbarians into assimilation: and this longstanding policy has been remarkably successful. Most peoples of the empire have been thoroughly Sinicized and now think of themselves as Chinese, whatever their ethnic origins. In consequence, the humanity of the barbarians was beyond doubt; and the Chinese elite never had to confront challenging or puzzling encounters with the unfamiliar 'savages' whom long-range commerce and adventure disclosed to European inspectors and specimen-hunters in modern times.

Challenges to assumptions about human nature therefore became a peculiar feature of European experience. By the thirteenth century, indeed, Europe was already on the threshold of an 'age of discovery' in which contacts with other cultures multiplied: the Mongol invasions of the thirteenth century; the improved communications which, in consequence, crossed Eurasia and linked Europe with China; the exploration of the African Atlantic and the discovery of a surprisingly 'primitive' culture in the Canary Islands in the fourteenth century; the accelerating contacts with black Africa in the fifteenth; the opening of the New World and of direct seaborne routes to the Indian Ocean from the 1490s; the traversal of the Pacific and its slow exploration in the early modern era; the huge accession of knowledge of new human cultures and non-human species which accompanied these events. The struggle was on to fit the new knowledge into the traditional classical, biblical, and folkloric panoramas of humankind. Meanwhile, Renaissance anatomists discovered that women were not merely nature's bodged attempts to make men; the long struggle unfolded to establish the fully human credentials of black people; and early-modern intellectuals wrestled with the problems posed by anatomical anomalies, such as those of pygmies and 'Hottentots'.

As Westerners' knowledge of the wider world increased, a check-list gradually developed of the criteria that could elevate a society to fully human status. Reason recognized natural law. Therefore the equipage of a society ruled by natural law was essential: government, laws, religion. There was intense debate from the thirteenth century onwards in Latin Christendom over how to recognize infractions of natural law, but there were some outrages—sodomy, bestiality, cannibalism, and human sacrifice—which most writers considered absolutely alienating; others, like blasphemy or unwillingness to listen to the gospel, were more controversial—unnatural to some, obviously cultural (though they did not put it like that) to others. The sexual prurience of late medieval and early-modern ethnographers is not, therefore, to be explained as submission to the seductions of pornography: it was a practical and scientific contribution to debate over whom to include in the human fold. Nor are horror-stories about cannibals and human sacrifice to be classed always as 'wonder-tales'—early-modern forms of tabloid journalism: they were included as responses to serious questions about the nature of the societies explorers encountered—questions, indeed, literally of life and death, since it was a common assumption of canon lawyers that adhesion to natural law was a qualification for the exercise of true sovereignty. Those who infringed natural law forfeited its protection and exposed themselves to just conquest by the right-minded.

These doctrines, formulated in the course of thirteenth-century debates about the proper attitude for Christendom to adopt towards pagan enemies, coloured the more remote encounters that followed. In a notorious sermon in 1344, Pope Clement VI summarized them in justifying his proclamation of what was, in effect, a crusade against the recently discovered Canary Islands, whose 'naked', pastoral inhabitants were enduring slaving *razzie* and other unwelcome attentions from European visitors. From the thirteenth century onwards, the importance of social criteria for identifying humankind was enhanced by the reception or re-absorption of Aristotle's *Politics* into the Western tradition. If man was 'by nature' political and social, then ungregarious habits would mark creatures out for exclusion.

At the nether edge of the heirarchy of societies were people who eschewed social life altogether. At least, such people were postulated, more, perhaps, from imagination than experience. The wild man of the woods, the *Homo silvestris*, was one of the many intermediary denizens of the woodlands that fringed the civilized West in the Middle Ages: beast-men, werewolves, vampires all came from occluded glades which stimulated imaginations in the forest-zone between humans and others. Wild men were favourite subjects of medieval art. They appeared in manuscript illuminations in the Rhineland, tapestries in the Low Countries, a painted ceiling in the Alhambra, a heraldic device in Normandy, tableware in Bavaria, and the carvings on a college doorway in Valladolid. They exhibited potential both for terror and taming. They abducted ladies, but then submitted to their captives, learning how to converse politely and play chess. War against the wild men—as against monsters and mythic beasts—was so common a theme of the representation of knightly activity that it is hard to resist the impression that the defence of civilization against savagery was a chivalric obligation. To meet a wild man's challenge, the hero of the fourteenth-century English poem, *Sir Gawain and the Green Knight*, had to cross a wilderness of 'hoary oaks', fighting off worms and wolves and treelike giants called 'entains'. He was 'near slain with sleet' among naked rocks where birds 'piped piteously'. His adversary was the colour of the forest, with hair like fronds and tree-like stature and solidity. Yet he had a touch of noble savagery about him and could teach morality to a knight of the round table.

Did wild men really exist? Sometimes, people thought they had found them. This is easily intelligible in the context of the time. In the Middle Ages the forest was literally the frontier of medieval Christendom: the terrain of expansion, the abode of paganism, the habitat of demons and old gods. When the prospects of expansion shifted to remoter frontiers, wild men became re-located in more expansive imaginations. It is no coincidence that images of wild men multiplied in the fifteenth and sixteenth centuries, at a time when European exploration was establishing contact with peoples who resembled them in the Canary Islands, Africa, and the New World: people who lived in 'wild' terrain, often densely forested, who practised social nakedness, or who in some cases went clad in pelts. A fifteenth-century illustrated manuscript of the *Roman de la Rose* shows a world of wild men and was clearly influenced by reports of recently discovered, cave-dwelling, hide-wearing Canarians. The coat of arms of one of the first conquistadores of the Canaries acquired supporters in the form of wild men. When he applied to Rome for bulls authorizing his violent raids on the coasts of west Africa, Henry the Navigator called the peoples who were his potential victims 'wild men of the woods'. At the French royal court in the sixteenth century, the roles of wild men were played by Brazilian Indians. Dating from about 1550, a typical entertainment, in which green, hairy men charge on stage to carry off the ladies, is depicted on the walls of a banqueting chamber in the Castle of Binches. A few years earlier, Jan Mostaert, court painter of Mary of Hungary, imagined a romanticized scene of warfare in the New World, re-casting the clash of natives and conquistadores in the traditional imagery of battles between wild men and knights. Wild men lived at the limits of what could reasonably be called a human way of life: a bit of domestication could establish them firmly in the ranks of humanity. A little slip, a smidgeon of degeneracy, could tip them down among the beasts.

Wild men formed a category of diminishing usefulness in a period of growing knowledge of human diversity. The peoples encountered during Europe's overseas 'expansion' were, on the whole, more like one another physically than medieval legend had predicted: most of the monsters, it turned out, did not really exist. On the other hand, the cultural variegation of humankind proved more intense than anyone could have supposed. There really were cannibals, for instance—something which Columbus, before he met them, dismissed as impossible—there were societies which practised every variety of sexual coupling, including nearly-free love; every sort of political heirarchy, including apparently near-perfect equality; every kind of community from primate-style bands to states of common allegiance which dwarfed those of Europe. The more 'primitive' a society seemed, the more interesting it was. Partly, this was simply because of the fascination of the unfamiliar; but it was also because of the presumption that present primitivism could illuminate the human past. Humanists' demand for information which could cast light on early humans' lives and language stimulated the quest for the primitive, and certainly prejudiced the terms in which explorers reported their finds.

The most exciting moment of all occurred when Columbus first glimpsed what he called 'naked people', on Friday, 12 October 1492, on an island he called 'San Salvador', which most scholars locate in the Bahamas. The natives were therefore probably Lucayos—a people of whom little is known, though archaeological evidence endorses Columbus's account of their rudimentary material culture. His description deploys many of the categories, analogies, and images available in his day to help Westerners understand other cultures. He compared the new-found people, implicitly or explicitly, with Canary Islanders, blacks, and 'monstrous', sub-human races: they were, he said, 'neither black nor white but like the Canarians', 'of goodly stature', and 'well proportioned'. The purpose of these comparisons was evidently not so much to convey an idea of what the islanders were like as to establish doctrinal points: the people were comparable with others who inhabited similar latitudes, in conformity with a doctrine of Aristotle's. They were physically normal and therefore—according to late-medieval psychology derived from Albertus Magnus—fully human and rational. This qualified them as potential converts to Christianity. Columbus went on to emphasize the natural goodness of these unwarlike innocents, uncorrupted by material greed—indeed, improved by poverty. They even had an inkling of natural religion undiverted into what were considered 'unnatural' channels such as idolatry. He emphasized that they went 'as naked as their mothers bore them'. Their nakedness suggested two kinds of innocence: Adam's and Eve's in Eden, and that of St Francis, who stripped himself naked at the start of his apostolate as a sign of total dependence on God (Columbus had many friends in the Franciscan Order and drew increasingly on Franciscan spirituality for comfort and guidance as his life went on). The way he depicted native society also recalled the 'Golden Age' of silvan innocence imagined by classical poets as prevailing in remote antiquity: the Renaissance revival of classical myth and learning made this image familiar in Columbus's day.

Finally, Columbus was alert for evidence that the natives were commercially exploitable as trading partners or slaves. At first sight, this seems at variance with his praise for their moral qualities; but many of his observations cut two ways. The natives' ignorance of warfare established their innocent credentials but also meant they would be easy to conquer. Their nakedness evoked a primitive idyll or an ideal of dependence on God, but also suggested savagery and similarity to beasts. Their commercial inexpertise showed that they were both morally uncorrupted and easily duped. Their rational faculties made them identifiable as human and exploitable as slaves. Columbus's attitude was not necessarily duplicitous, only ambiguous: he seemed genuinely torn between conflicting perceptions. After all, he and his men were undergoing an experience no European had ever had before.

Within the hurried time-frame of his first voyage around the Caribbean, the fragmentary narrative Columbus left us shows how his mind reacted and adapted to what he saw, as he struggled to accommodate new observations into the assumptions and prejudices with which he started. Towards the end of 1492, he was exploring the shores of Hispaniola, where he found himself among Arawak peoples, whose material culture was more impressive, by European standards, than that of peoples previously encountered. Their artefacts included elaborate stonework and woodwork in ceremonial spaces, stone collars and pendants and richly carved thrones. Columbus began to re-cast the natives as potential trading partners and mediators with the great civilizations of Asia, which, he hoped, lay only a short distance away by sea. Throughout his journeys to the New World Columbus remained undecided between rival perceptions of the people—as potential Christians, as exemplars of pagan virtue, as exploitable chattels, as savage, as civilized, as figures of fun.

European eyes adjusted to the realities of cultural diversity. The old topoi of wild men and the check-list of evidence of adherence to natural law became outmoded. Influenced by missionaries eager to save souls among newly encountered peoples, the Church took a positive view of their natural qualities in an effort to protect them from secular depredations, exploitation, and extermination. The question of whether the native peoples of the New World were fully human, endowed with rational souls, was settled by Pope Paul III in the 1530s, but their status needed frequent shoring-up against slippage. Missionary ethnographers in the Americas laboriously built up dossiers to demonstrate the social and political sophistication of native societies. A case like that of the Aztecs posed typical problems: cannibalism and human sacrifice besmirched the record of a people who otherwise appeared highly 'civil'; in evidence—vividly painted by native artists at the court of the Viceroy of Mexico and compiled under missionary guidance in the 1540s—one can still see the range of qualities the clergy held up for admiration. The training of an Aztec oblate is shown in gory detail, as his teachers beat his body to bleeding: this was presented as evidence not of barbarism but of the similarity of Aztec values to those of their Franciscan evangelists, who also practised devotional flagellation and tortured their flesh in mortification. The Aztec polity was depicted as a well-regulated pyramid, symmetrically disposed for the administration of justice, with an emperor at the top, counsellors below him, and common supplicants at the lowest level: a mirror-image of the society the missionaries had left back home. The Aztecs' sense of justice was shown to conform to the rather self-reflexive standards Europeans deemed 'natural': an adulterous couple, stoned to death, suggested an analogy with the ancient Jews and, therefore, prospective receptivity to the milder Christian message. Justice was tempered with mercy: though drunkenness was punishable by death, the aged were depicted as enjoying the exemption of mild restraint. When the Aztecs went to war, provocations on their enemies' part were shown to precede hostilities, which followed only after diplomatic efforts had been rebuffed. The natives, it seemed, practised just war by traditional Christian criteria—something which the Spanish monarchy strove to do with imperfect success. Examples like these could be cited for every native community where missionaries worked.

Bartolomé de Las Casas was the loudest spokesman for an inclusive attitude to the definition of humankind. He was a convert to conscience—a reformed exploiter of Indian labour on Hispaniola, who reformed in 1511 when he heard a Dominican preacher's challenge: 'are the Indians not human beings, endowed with rational souls, like yourselves?' He joined the Dominicans and became the crown's officially appointed 'Protector of the Indians': in effect, despite unsuccessful spells as a missionary and a frontier bishop, he was a professional lobbyist who managed, albeit briefly, to get the Spanish monarchy to legislate for Indian rights. Human sacrifice, according to Las Casas, should be seen rather as evidence of the misplaced piety of its practitioners, or of their pitiable state as victims of diabolic delusion, than as an infringment of natural law. His conclusion—'All the peoples of mankind are human'—sounds like a tautology; but it was a message important enough to bear repetition. Even cannibalism could be re-classified as an historical relic rather than an unnatural perversion—evidence of a primitive stage of social development, which all societies went through. Las Casas argued this with great flair in a work of the 1550s which was too long to be published even in those expansive days; but in 1580 Montaigne produced a pithy, elegant, and famous defence of cannibalism which also included a reproach against his own compatriots, whose humanity to each other took, he thought, different but at least equally evil forms.

Montaigne seems to have felt that a 'natural' vice was better than a 'civilized' one, which was tainted by contrivance and unexcused by ignorance. Correspondingly, when it came to virtues, the savage were again better than the civilized. The logical conclusion of this line of thought was the doctrine of the 'good' or 'noble' savage, whose natural goodness was unalloyed by convention, unstaled by custom, uncorrupted by interest. The doctrine made increasing sense in the seventeenth and eighteenth centuries in Europe, when optimistic, even Panglossian accounts of human nature were common; on the other hand, it ran counter to the dogma of progress and provoked scepticism from two still-influential schools: orthodox theologians, who upheld the effects of original sin; and political reactionaries, who were convinced that people needed strong rulers to hold their evil instincts in check. As a result, the appeal of the noble savage was predictable: radicals and anticlericals loved him and tended to believe in him. As the period lengthened, romanticism allied with primitivism to give him a further constituency.

The original 'noble savage', explicitly so called, was a Micmac Indian of the Canadian woodland, described by Marc Lescarbot, who spent a couple of years in Nouvelle France in the early seventeenth century. He regarded the

Micmac as 'truly noble' in the strictest sense of the word, because their menfolk practised the noble occupations of hunting and arms. But they also exhibited virtues that civilization corroded: generosity ('this mutual charity which we seem to have lost'), a natural sense of law ('so they have quarrels very seldom'), common life and property. Ambition and corruption were unknown among them. But this was an imperfect Eden, where violence was often vindictive and austerity unknown in meat and drink. Nor did Lescarbot's admiration for Micmac morality make him less inclined to justify conquering them and depriving them of their sovereignty and their land.

The idea of the noble savage really became rooted in Western tradition when it was transferred to the Huron. Redemptorist and Jesuit missionaries were repelled by some of the culture they found on the banks of the Great Lakes—especially the horrifying rituals of human sacrifice, in which captives were tortured to death for days on end. Among Iroquoian peoples, however, they were unable to resist a distinct partiality for the Huron because the latter were exceptionally welcoming to them and responsive to their Christian message. Of course, it was not a disinterested reception—the Huron were usually at war with their neighbours and desperate for allies—but the missionaries felt its warmth. The very first of them, Gaspard Sagard, who visited the Huron in 1623, was the founder of what could be called Huronophilia with his selective praise for their kindnesses to him and to each other, their egalitarian values and the technical proficiency, as builders, farmers, and canoewrights, which, he thought, made them superior to the Algonquians to their east. They even had a system of glyphs which demonstrated their ascent to literacy: indeed, Iroquoian peoples did record topographical data and the outcomes of battles with symbolic annotations carved on tree trunks.

Although missionaries were candid in their criticisms of the defects of the savage way of life, the secular philosophers who read them tended to accentuate the positive and eliminate the negative. Cautionary tales were filtered out of the missionary relations and only an idealized Huron remained. This transformation of tradition into legend became easier as real Huron literally disappeared—first decimated, then virtually destroyed by the diseases to which European contagion exposed them.

The great secularizer of Huronophilia was Louis-Armand de Lom de l'Arce, who called himself by the title his family had sold for cash, 'Sieur de Lahontan'. Like many refugees from a world of restricted social opportunity at home, he went to Canada in the 1680s and set himself up as an expert on its curiosities. The mouthpiece for his free-thinking anticlericalism was an invented Huron interlocutor called Adario, with whom he walked in the woods, discussing the imperfections of biblical translations, the virtues of republicanism, and the merits of free love. His devastating satire on the Church, the monarchy, and the pretensions and pettiness of the French *haut monde* fed directly into

Voltaire's tale of the 1760s of an 'ingenuous' Huron sage in Paris.

The socially inebriating potential of the Huron myth was distilled in a comedy of uncertain authorship, performed in Paris in 1768, which also inspired or plagiarized Voltaire's portrait. The Huron excels in all the virtues of noble savagery as huntsman, lover, and warrior against the English. He traverses the world with an intellectual's ambition: 'to see a little of how it is made'. When urged to adopt French dress he denounces imitation as fashion 'among monkeys but not among men'. 'If he lacks enlightenment by great minds,' opines an observer, 'he has abundant sentiments, which I esteem more highly. And I fear that in becoming civilized he will be the poorer.' Victimized by a typical love-triangle of the comedy of manners, the Huron exhorts the mob to breach the Bastile to rescue his imprisoned love. He is therefore arrested for sedition. 'His crime is manifest. It is an uprising.' This seems a remarkable prefiguration of 1789.

The stock of images of noble savagery was topped up in the late eighteenth century by exploration of the Pacific and the specimens of Pacific manhood the explorers brought home. In 1774, English society lionized Omai, who had been a restless misfit in his native Polynesia. Duchesses praised his natural graces and Reynolds painted him as a type of equipoise and uncorrupted dignity. Lee Boo, from Palau in Micronesia, was equally convincing as a 'prince of nature'. Visitors to the Pacific found a voluptuary's paradise.

Bougainville called Tahiti the 'Isle of Cythera' and the ease with which sexual favours could be obtained from native women became one of the most persistent topoi of literature about the place. Romantic primitivism became inseparable from sexual opportunity. Lahontan had established the connection, with his recommendations of the uncomplicated connubiality of Huron mating customs, illustrated with engravings of women signifying their consent by blowing on torches carried by the partners of their choice. Now images of Tahiti as the ravishing habitat of inviting nymphs filled Westerners' canvases, from William Hodges—the illustrator of Captain Cook's voyage—to Gauguin. The sensuality of primitivism clung to less likely climes. Philosophical idealization of the Sami, which began in the eighteenth century, was lubricated by the sweat of the 'promiscuously' mixed-sex sauna.

The last echoes of the hunt for the wild man sounded in the eighteenth century. The disappointments of previous centuries had not allayed the quest for 'natural' man. On the contrary, interest in such problems as the origins of language, the origins of political and social life, and the moral effects of civilization was never so acute. Savants' anxiety to examine specimens of unsocialized primitivism was greater than ever. 'Noble savages' were brought from the extremities of empire—from the Great Lakes of North America and the islands of the South Seas—for exhibition and examination in London and Paris, but even they represented too advanced a phase of the development of society to satisfy scientific

16. Cesare Lombroso (1835-1909) adapted craniology, phrenology, and physiognomy to the attempt to identify criminal types, who, he claimed, represented throwbacks to primitive stages of human evolution. The evidence consisted in what he called 'stigmata' – abnormalities in jawlines and facial proportions.

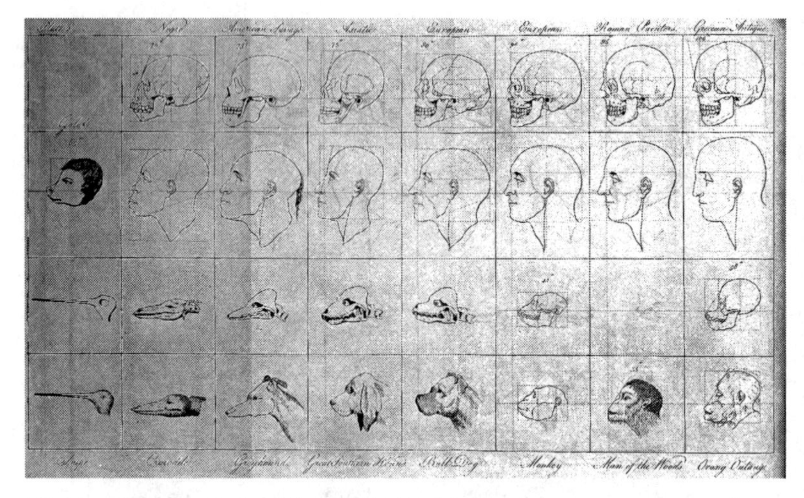

17. The anatomist Charles White (1728-1813) believed that 'various species of men were originally created and separated by marks sufficiently discriminative' to exhibit their place in a heirarchy of nature, with white people 'most removed from brute creation', while the bodies and, especially, the skulls of blacks 'differed from the European and approached to the ape.'

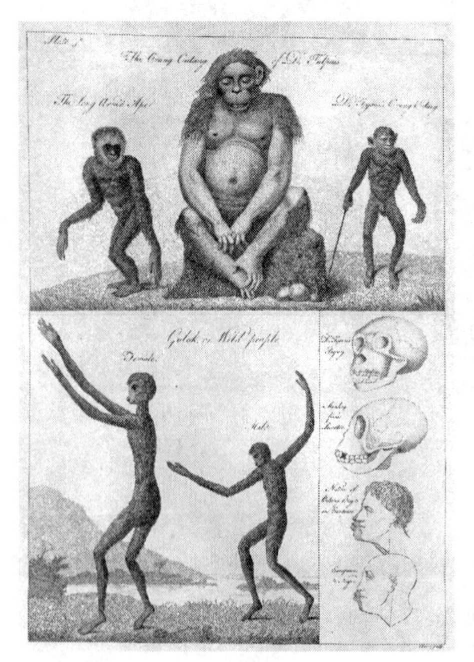

18. 'This plate,' White wrote, 'exhibits copies of the best authenticated engraving … of Apes, which approach nearest to Man: likewise the skull of Dr Tyson's pygmy – the skull of a monkey, … the profiles of a native of Botany Bay and an European – and profiles of an African and an European.' Tulpius's orang-utan and Tyson's 'orang-utan' (actually a chimp) are well reproduced (see p. 84). Notice the chimp's stick – a longstanding iconographical commonplace – and long penis, which White cited as evidence of the proximity of apes to blacks.

19. Wild men, or wearers of wild-man disguise, in mock jousts – jokey or fantastic – are plentiful in the illustrations of the Hours of Engelbert of Nassau, one of the last great illuminated prayer books, by the Master of Mary of Burgundy. But real conflicts between knights and wild men are also common in the genre: fighting to the death, or exchanging and sometimes executing captives.

20. In many places in medieval western Europe, the Wild Man symbolized carnival, with its relaxations of standards of civilized behaviour, and was symbolically killed and 'buried' at its conclusion. In the early sixteenth century, Nuremberg's famous Schembartläufer – with their traditional right to beat spectators – represented the myth of the 'Wild Horde' of demons, who wasted the winter countryside, and whose ritual defeat celebrated the promise of spring.

21. Hans Staden's sensational account of his captivity among cannibals in Brazil appeared in 1557. De Bry's engravings for the 1592 edition became more famous and influential than the text. The victim's entrails, Staden explains, 'are kept by the women who boil them and make a thick broth. This they and the children drink. They devour the bowels and flesh from the head.'

22. Jan Mostaert's account of a conquistador encounter was painted in the 1540s, probably of Mary of Burgundy. Presumably intended to celebrate the achievements of the House of Habsburg, it is highly equivocal. The conquerors interrupt an idyllic scene of naked people – suggestive of Edenic innocence or dependence on God – with romantic landscapes and Arcadian pastures. The peaceable natives have no weapons but fight back with rocks and sticks.

curiosity. 'Wolf-children' seemed, for a while, to be likely to supply the required raw material for analysis. Linnaeus supposed they were a distinct species of the genus *Homo*—*Homo ferens*, embodiments of a wild-man myth which turned out to be true. Plucked from whatever woods they were found in, wrenched from the dugs of vulpine surrogate-mothers, they became experiments in civilization, subjected to efforts to teach them language and manners.

Numbers of recorded cases quicken in the seventeenth and eighteenth centuries. Was this because of renewed interest in feral children, stimulated by analogies with the 'savages' enumerated by overseas expansion? Or was it simply a function of the explosion of population in the Europe of the day, expanding the limits of towns and cultivation, squeezing the remaining tracts of unpopulated 'wilderness'? All the experiments failed. Boys supposedly raised by bears in seventeenth-century Poland continued to prefer the company of bears. 'Peter the Wild Boy' whom rival members of the English royal family struggled to possess as a pet in the 1720s, hated clothes and beds and never learned to talk. The 'savage girl' kidnapped from the woods near Songi in 1731 preferred fresh frogs to the viands of the kitchen of the Chateau d'Epinoy and was for a long time more adept in imitating birdsong than speaking French. The most famous case of all was that of the 'Wild Boy of Aveyron'. Abandoned in infancy in the high forest of the Tarn, he survived by his own wits for years until he was kidnapped for civilization in 1798. He learned to wear clothes and to dine elegantly, but never to speak or to like what had happened to him. His tutor described him drinking fastidiously after dinner in the window of the room, 'as if in this moment of happiness this child of nature tries to unite the only two good things which have survived the loss of his liberty—a drink of limpid water and the sight of sun and country'.

Darwin himself witnessed and participated in the last, equally unsatisfactory attempt to domesticate people captured from the wild. Among fellow voyagers aboard the *Beagle*, which took him on his voyage around the world in 1831–6, were three Fuegians whom a previous British expedition had seized as hostages and transported to England, to be treated as the philosophes of the previous century had treated the 'wolf-children'. They were taught English and Christianity and the refinements of etiquette. They were dressed and groomed. One of them, known as Jemmy Button after the pearl button supposedly given to his parents to compensate for his kidnap, became a notorious dandy, who wore gloves and got upset if his shoes were dirty. The others were a betrothed couple, called Fuegia Basket and York Minster, who were married on arrival in Tierra del Fuego by the missionary who accompanied the party, Robert Mathews. In theory, the wanderers' return to the wild would present the Fuegian 'savages' with knowledge and a model of civilized life and precede their conversion and domestication. Man might be suckled by wolves, but his destiny was to found Rome.

The experiment began to go wrong even before the *Beagle* had left the shore. The natives showed no respect for the returnees' transformation, and little interest in communicating with them. They plundered their goods and drove Mathews distracted with their depredations and threats of violence, so that he had to be taken on board again. Button was uncomfortable and ashamed of his relatives, affecting a white man's exasperation with the natives' ignorance and brutality. The crew of the ship left him there with misgivings. By the time the *Beagle* returned fifteen months later, the returnees had reverted to the lifeways of the wild, sliding back into their old relationships and joining in the fighting and plunder of their tribes. Jemmy Button was still friendly to his former shipmates, but they found him physically re-transformed: naked, dishevelled, dirty, wiry, and warlike. In future years he became a leader of native resistance against missionaries and an instigator of massacre. York Minster was killed in an inter-tribal brawl. Fuegia Basket lived until 1883: when last seen by a missionary, she had forgotten England, English, Christianity, and everything about her role in Fitzroy's doomed experiment.

By Darwin's day, the scientific world more or less united in rejecting Lord Monboddo's theory that orangutans were human. By that time, the configurations of humankind were more or less as we now think them to be, with none of the exclusions which had dappled earlier discussions. But the problem was cast back into the crucible by nineteenth-century developments. Scientific racism multiplied the sub-categories into which humankind was split. The new science of social anthropology proposed cultural as well as biological criteria of differentiation. John Lubbock, for instance, was the Richard Dawkins of his day. Lubbock was Darwin's neighbour in Kent, and one of his earliest and closest adherents, who set himself up as an 'expositor of science' and 'mentor to the general public'. Of all the great range of polymathic works with which he piled the bookshop-shelves, none was more influential than *Prehistoric Times* (1865), in which he propounded a cultural counterpart of the theory of evolution: Tasmanians and Fuegians were 'to the antiquary what the opossum and the sloth are' to biologists: throwbacks to an earlier phase, living evidence—albeit doomed to extinction—of the antiquity of humankind and of the savagery of archaic humans. Ethnographers attracted to the study of 'primitives' felt they were journeying into the past. Cultural anthropology arose—it is tempting to suspect—in a partisan response to the secularization of science: as it became ever harder to invoke metaphysics in favour of the special nature of humanhood—harder to invoke God or cite the soul—culture became an alternative, secular, scientifically verifiable differentiator: a secular soul, something only humans had. In the long run, of course, as we have seen, this proved to be a false assumption.

Nineteenth-century imperialism cloyed European appetites for savagery. When romantic primitivism revived in the twentieth century, sex was the spur and civilized repression the enemy. The noble savage resurfaced among anthropologists drawn back to the Pacific like the lovestruck mariners of the age of Enlightenment. Margaret Mead's *Coming of Age in Samoa*, published in 1928, was based on fieldwork with pubescent girls in a supposedly unrepressive society. Whether the paradise Mead depicted was factual or fantastic has been much debated. The image, however, was seductive. Now that no one believes in the survival of Eden, romantic primitivism has taken refuge on remoter frontiers: in prehistory, or among those frolicsome simians, the bonobos or pygmy chimpanzees of the Congo. These 'apes from Venus', discovered in 1929, 'make love not war'. It is true that no case of inter-communal warfare, such as chimpanzees practise, has yet been detected among bonobos. Within their communities violent competition over food and mating is much less than among chimpanzees, but is by no means entirely absent. Their sexual enthusiasm is beyond comparison. According to studies by Frans de Waal, they 'engage in sex in virtually every partner combination' and 'every imaginable position and variation'. The facts that females exercise dominance over males and exhibit a preference for each others' company has also made bonobos heroines of feminism.

So the effort to erect a cultural threshold for admission to human status has failed to cope with two problems: human cultural diversity, which makes universal features hard to identify; and non-human animal culture, which undermines human claims to exclusive proprietary rights in culture. Meanwhile, a deadlier intellectual virus has been at work on the concept of humankind: the theory of evolution, which, by locating humans in an animal continuum in which there are no well-defined boundaries, created a new obstacle to the development of a discrete notion of our nature. Science repeatedly draws us back to awareness of the continuities which link us to the rest of creation. The next chapter is about the consequences.

CHAPTER 4

THE EVOLUTIONARY PREDICAMENT

Confrontations with Hominids

'Bare, Fork'd Animal': Encountering Unaccommodated Man

Orang-utans, whose influence on humans' self-image has been so pervasive, were a further source of inspiration for Charles Darwin. He liked to visit London Zoo to observe little Jenny, the menagerie's curious specimen of the species. She was, he thought, uncannily like a human child, understanding her keeper's language, wheedling treats, and showing off her pretty dress when her keepers presented her to the Duchess of Cambridge. Darwin evidently preferred her to some of the humans he knew. In particular, he found the natives of Tierra del Fuego repulsive when he first saw them aboard the *Beagle* in 1832: 'man in his lowest state,' they seemed to him, apparently 'bereft of human reason or at least of arts consequent to that reason.' The sight convinced him of a thought so terrible that he did not even dare to confide it to *The Origin of Species*: man was an animal like other animals. 'The difference between savage and civilised man,' he added, 'is greater than between a wild and domesticated animal.' Islanders' language 'scarcely deserves to be considered articulate. Captain Cook has compared it to a man clearing his throat, but certainly no European ever cleared his throat with so many hoarse, guttural and clicking sounds.' The specimens encountered later in the voyage, on the western side of the island, were even more bestial, sleeping 'coiled up like animals on the wet ground', condemned by cold and poverty to a life of 'famine, and, as a consequence, cannibalism accompanied by patricide',

> stunted in their growth, their hideous faces bedaubed with white paint, their skins filthy and greasy, their hair entangled, their gestures violent and without dignity. Viewing such men, one can hardly make oneself believe they are fellow-creatures and inhabitants of the same world. . . . How little can the higher powers of the mind be brought into play! What is there for imagination to picture, for reason to compare, for judgement to decide upon? To knock a limpet from the rock does not even require cunning, that lowest power of the mind. Their skill in some respects may be compared to the instinct of animals; for it is not improved by experience.

Part of the germ of the Theory of Evolution entered his head as he puzzled over how the Fuegians could endure the climate in a state of near-nakedness. 'Nature, by making habit omnipotent and its effects hereditary, has fitted the Fuegian to the climate and productions of his miserable country.' The narrative of the genealogy of man, which Darwin published in 1871, started with marine animalculi which he likened to larvae. From these descended fish, from whom 'a very small advance would carry us on to the amphibians . . . but no one can at present say by what line of descent the . . . mammals, birds and reptiles were derived from . . . amphibians and fishes'. Among mammals, placental animals succeeded marsupials.

> We may thus ascend to the Lemuridae; and the interval is not wide from these to the Simiadae. The Simiadae then branched off into two great stems, the New World and Old World monkeys; and from the latter, at a remote period, Man, the wonder and glory of the Universe, proceeded. Thus we have given to man a pedigree of prodigious length, but not, it may be said, of noble quality.

We thus learn that 'man is descended from a hairy quadruped, furnished with a tail and pointed ears, probably arboreal in its habits . . .'

Now here is a dazzling irony: tradition conceived of apes as degenerate men. Darwin re-classed men as evolved apes. Within a couple of generations of the rejection of Lord Monboddo's theory, the problem of defining the limits of humankind had been flung back into the crucible. The orang-utan could reclaim his lost place as a 'brother' or, at least, a cousin of humankind. How should our common ancestors be classed: as human, or as non-human animals? The problem gained urgency at about the time Darwin wrote. The first Neanderthal remains turned up in Spain in 1848; the next find—which led directly to the identification of the species—occurred in the Neander Valley three years before the publication of *The Origin of Species*. Thereafter, new discoveries happened frequently—partly because diggers were on the look-out for them. The human family tree sprouted budding ancestors, as well as growing limbs and branches which led in other directions.

Votes for Oysters?

The theory of evolution is obviously—broadly speaking—true. Darwin was not right about all the details or even all of the main elements of the theory. Natural selection does not seem able to account for all the differences between species without allowing some scope for random mutations. Evolution has probably not unfolded at the continuous pace postulated in Darwin's day, but seems to have resembled a state of equilibrium punctuated by spasms and lurches. The conditions and environments in which life began may have produced far more original organisms—and therefore far more many lines of descent to the biota which populate today's biosphere—than he supposed. Evolutionary theory as he conceived it resembles a religion: evolution functions as a sort of ersatz-Providence, determining what happens to everything on Earth; it is more convincing—more characteristic of what we now know about our planet—if it is re-cast as a chaotic system, full of untraceable paths, untrackable causes, and unpredictable effects. Still, the basic model is rationally unchallengeable: all biota belong to the same evolutionary continuum. Humans, in particular, share common ancestry with other animals and—by evolutionary standards—close common ancestry with other apes. In the context of this model, is humankind a coherent concept? At what point in the history of evolution might it make sense to distinguish humans from non-humans? Bertrand Russell, with the relentless clear-headedness of a great logician, declared the distinction unsustainable: evolutionary theory demanded the re-drawing of our frontier with the animal kingdom, to which there was no logical conclusion short of 'votes for oysters'. Darwin himself was fully aware of this inescapable sequence of thoughts. 'In a series of forms,' he wrote in *The Descent of Man*, 'graduating insensibly from some ape-like creature to man as he now exists, it would be impossible to fix on any definite point when the term "man" ought to be used.' He concluded, however, that this was 'a matter of very little importance'.

He meant, I think, that it was a purely terminological matter. Yet at moments he was inclined to submerge the question in broader ruminations on the supposedly spirtual nature of man. 'He who believes in the advancement of man from some lowly-organised form, will naturally ask how does this bear on the belief in the immortality of the soul.' This, too, was of very little importance in Darwin's scheme.

> Few persons feel any anxiety from the impossibility of determining at what precise period in the development of the individual, from the first trace of the minute germinal vesicle to the child either before or after birth, man becomes an immortal being; and there is no greater cause for anxiety because the period in the gradually ascending organic scale cannot possibly be determined.

He thought the same about the question of whether 'the so-called races' should be 'ranked as species or as sub-species'. Since man 'attained the rank of manhood, he has diverged into distinct races, or as they may more appropriately be called, sub-species. Some of these, for instance the Negro and European, are so distinct that, if specimens had been brought to a naturalist without any further information, they would undoubtedly have been considered by him as good and true species.' Most debate has shied from such a radical proposal and centred on the vast twilight zone of human evolution dominated by the species we loosely call 'hominids'—the primates in the line of descent of *Homo sapiens*. At what point in that line of descent can we rationally begin to distinguish human from non-human beings? An eleven-year-old boy who died—or was left after death—in the 'Cave of Bones' of Atapuerca 800,000 years ago, had a facial structure much more like ours than that of later-evolved species. And what reasonable grounds can there be for excluding primates in parallel lines of descent, characterized by close resemblances to our own? These questions are inextricably linked but, for convenience, we can take them in turn.

Traditionally, palaeoanthropologists have drawn a dividing line at the emergence in the fossil record of a creature known as *Homo habilis*, about two and a half million years ago. This is the first species to be credited with the generic name 'Homo'. Earlier hominids are classified, with very different resonance, as 'pithecanthropoi'—literally, apemen, or, in current terminology, 'australopithecines'—literally, 'southern ape-like beings'. The distinctive thing about *habilis* was his toolkit: he chipped hand-axes from stones. By privileging him, therefore, scholars aligned themselves with a now old-fashioned—indeed, discredited—definition of man as 'tool-maker'. A slightly later variant, *Homo ergaster*, gets some experts' vote on the grounds that he stacked the bones of his dead, and therefore exhibited some form of 'higher consciousness' concerned with more than a competitive advantage in the struggle for life. Alternatively, *Homo erectus* is preferred because his neatly napped, symmetrical flints appeal to excavators' aesthetic sensibilities. Clearly, all these instances of the championship of one set of ancestors over another reflect subjective criteria: the qualifications demanded are a matter of resemblance to ourselves.

In selective respects, we seem to be willing to venerate any ancestor, however remote. The world seems to have warmed—that really is, I think, the right word—to a creature mentally reconstructed from fossils over three million years old in Afar in Ethiopia. Don Johanson found her bones in November 1974, identified her as a bipedal with an ape-like brain, and called her 'Lucy' after the title of a song on a Beatles record he played the same evening in his camp. Subsequent finds suggested she was a member of a thirteen-strong family and the area she lived in was occupied less than half a million years later by tool-makers. The find forced back in time the threshold of developments which the world thought were defining characteristics of humankind: walking bipedally, wielding tools. Ancestors hundreds of thousands of years earlier than any to whom we concede the label 'Homo' anticipated our achievements in these respects. Lucy seemed to be the 'African Eve' the headline-writers longed for. (The same headline announced, with greater justice, various later attempts to use DNA evidence to identify the last common female ancestor of all living humans: she probably lived in Africa 143,000 years ago.)

Lucy was a triumph of PR. Separated from us by over three million years, and by vast chasms of physical, mental, and cultural differences, she seems to symbolize the flexibility of our modern concept of humankind. If we can accept Lucy as an ancestress, it helps to stretch the elasticity of the embrace in which we clasp each other, regardless of colour or creed, or outward appearance or mental resources or moral worth. This is a commendable but conceptually problematic temptation. Here is a useful test: would those of us who happen to be black, for instance, wish to admit to feeling a closer sense of community with some black person who had not yet emerged 'out of Africa'—say, forty or fifty thousand years ago—than with white neighbours, with whom our common ancestry was very much more remote but whom, in most respects, we closely resemble and with whom we share so much? If we were to do so, would we not be self-condemned as racist? In other words, why choose Lucy as 'the first human'? Why exclude earlier creatures of common ancestry or—even more to the point—later ones with whom we feel less affinity? If, for example, we privilege Lucy as human but exclude other primates who are our neighbours today or who were our neighbours in the past, are we not guilty of the same sort of prejudice as racists? There are other primates in the panorama of evolution who resemble us more closely than Lucy, and share her—or creatures very like her—among their common ancestors; yet we are curiously reluctant to admit them to human status. If, for example, like a now-discredited past generation of scholars, we wish to privilege bipedalism as a criterion of humanhood, we should have to include *Ardepithecus ramidus*, an otherwise apelike hominid who was capable of two-legged locomotion 4.4 million years ago but was still at home in trees. *Australopithecus*, whose earliest fossil evidence dates from two or three hundred thousand years later and who endured for three million years, achieved another much-vaunted criterion: he had hands capable of manipulating tools. *Homo habilis*, who overlapped with him, had apparently unobjectionable credentials: he made the earliest surviving stone tools we know of and had the 'big brain' beloved of nineteenth-century craniologists—half as big again as those of his predecessors.

All these examples show the difficulty of drawing a satisfactory line between human and non-human hominids. The key example, however, is that of the Neanderthals. Save for an accident of evolution, this species might still be around to challenge our human sense of uniqueness. If one were to meet one in the street, one might experience the same sense of kinship, of instant recognition across differences of aspect and problems of communication, typical of encounters between enlightened modern humans of different races or cultures. Neanderthals looked rather like us and behaved rather as ancestors of our own species did. To judge from specimens excavated so far, they had, if anything, bigger brains than *Homo sapiens*, comparable minds, and highly similar forms of culture. A Neanderthal family is buried together at La Ferrassie: two adults of different sexes are curled into the foetal position characteristic of Neanderthal burials all over what are now Europe and the Near East. Nearby, three children of between three and five years old and a newborn baby lie with flint tools and fragments of animal bones. The remains of an undeveloped foetus, extracted from the womb, is interred with the same dignity as the other family members, albeit without the tools. Other Neanderthal burials have more valuable grave goods: a pair of ibex horns accompanied one youth in death, a sprinkling of ochre was strewn on another. At Shanidar, in what is now Iraq, an old man—who had survived in the care of his community for many years after the loss of the use of an arm, severe disablement to both legs, and blindness in one eye—lies with traces of flowers and medicinal herbs.

All these cases—and many others of what look like ritual Neanderthal burials—have been challenged by sceptical scholars: 'explained away' as the results of accident or fraud. But there are too many of them to discount. In an effort to exclude Neanderthals from a share of humanity, sceptics have displayed apelike agility in challenging the facts: floods 'must' or 'might' have carried the remains to apparent burial sites. Lions or hyenas dragged corpses, complete with the beds they lay on. A sleeping Neanderthal in Shanidar 'must have' buried himself by accident by dislodging the cave ceiling. What look like floral offerings must have been blown to their resting-place by a chance breeze. At the other extreme, credulity has drawn irresponsible inferences from this evidence, crediting Neanderthals with a broad concept of humanity, a belief in the immortality of the soul, a system of social welfare, a gerontocracy, a political system of philosopher-rule. They may have had such things: but the burials are not evidence of them.

So what do the Neanderthal burials prove? Mere burial is evidence only of material concerns: to deter scavengers, to mask the odour of putrescence. But ritual burial is evidence of an idea: indeed of two ideas—of life and death. We still find it hard to define them and in particular cases—such as impenetrable comas and the misery of the moribund on life-support—to say exactly where the difference between them lies. But the conceptual distinction we make between them goes back thirty or forty thousand years, when people began to mark it by rites of differentiation of the dead. These first celebrations of death hallowed life. They constitute the first evidence of a more than merely instinctive valuing of life: a conviction that life is worthy of reverence, which has remained the basis of all human moral action ever since. They were part—perhaps the most eloquent part—of the common culture Neanderthals shared with *Homo sapiens*; some of the present survivors of *Homo sapiens*, however, seem unwilling to accept that. The tenacity with which some scientists strive to confine the Neanderthals to a 'lower' order of creation is so virulent, so committed, so obstinate, so indifferent to facts, that it is unintelligible except as part of a partisan programme with an ideological agenda.

The claim, for instance, that Neanderthals were incapable of language has now been disproved by the discovery of a fragment of a Neanderthal larynx at Kebara in what is now Israel: the sounds of Neanderthal vocalizations would have been different from ours, but they were capable of a range fully adequate to produce language in a sense analogous to that of 'modern' human speech. What is remarkable, however, is that the debate about Neanderthal language should ever have been conducted at such a superficial level: the assumption that some merely physical impediment would be sufficient to frustrate the development of language displays a fundamental ignorance of what language is. It would mean that the symbolic forms of expression used in sign language would not qualify. Or that morse code, which uses only one sound, was incapable of expressing the same range of meanings as—say—English. Obviously, anyone prepared to concede to Neanderthals a level and form of intelligence comparable with our own would suppose them capable of devising a language consistent with the limitations and opportunities of their vocal tracts.

Equally unconvincing is the claim that Neanderthals were inexpert hunters—'opportunists', like hunting chimpanzees, who exploited prey in their immediate environs–whereas *Homo sapiens* is a prescient planner, who tracks his victims, plots their movements, and links them to the passage of the seasons and the state of the ecosytem. This distinction seems unlikely to be valid, as hominids of various kinds had all the cognitive apparatus necessary for planned hunting—to say nothing of appropriate technology, represented by cunningly flaked flints and spear-throwers—long before the evolution of the Neanderthals, whose material culture was, in other respects, so impressive for its time. More generally, the Neanderthals' detractors say that their extinction was the result of failure to adapt: but they colonized Europe as far as 52 degrees N—a habitat which was extreme even in the interglacial area when the Neanderthals penetrated it. Their subsequent retreat from those latitudes, as the Ice Age encroached, and their replacement by our ancestors, is better explained by the rhythms of the Ice Age—climate change which outstripped the responses even of these highly adaptable people—than by any doctrine of their inherent inferiority.

It is also commonly said that the Neanderthals were self-excluded from humankind by their lack of art. It is true that they left little art that has survived; but neither did our own ancestors until a date more or less equivalent to that of the Neanderthal extinction. In any case, there is one Neanderthal site in Europe—the Cave of the Reindeer at Arcy-sur-Cure—where the remains of necklaces of beads and ivory have been found. Denigrators of the Neanderthals ascribe these to the influence of (or commerce with) *Homo sapiens* but must at least admit that it shows the Neanderthals could share, if not originate, culture which, according to earlier orthodoxy, was peculiarly the product of our own ancestors. This find is also rich in red ochre—a substance usually associated with body-painting and the only evidence of this otherwise highly perishable form of early art.

Those who want to exclude Neanderthals from the human community altogether, or relegate them to an inferior class of humankind, argue that their capacities were inferior, their culture underdeveloped, their intelligence limited, their rational faculties defective, and that their potential for inter-breeding with properly human primates was vitiated by unbridgeable incompatibilities. The survival of aged and crippled Neanderthals means nothing: a famous monkey called Mozu survived to an advanced age in the Shiga Heights, despite being abominably crippled, without much help from her tribe. Neanderthals' early extinction is often treated as evidence of their inferiority, but they lasted for 300,000 years: much longer than *Homo sapiens* has managed so far or, on present showing, looks like managing in future. Readers of this book will recognize what is at stake. The terms in which modern palaeoanthropologists debate the nature of the Neanderthals are strikingly—frighteningly—similar to those of nineteenth-century debates about blacks.

The evidence supports a dispassionate conclusion: there have been non-human species in the past who were morally indistiguishable from humans: Neanderthals—and lots of other hominids, who shared everything which to us seems essential to human nature. This is an important conclusion to bear in mind when considering whether the species *Homo sapiens* constitutes a coherent moral category. If we can embrace other hominid species in our circle of recognition, and acknowledge in them the essential qualities we ascribe to ourselves, can we exclude them from the same moral category? If we did so, would it be any more justifiable than excluding other races? These questions raise another: what, if any, is the moral difference between racism and species-ism? The notion of meeting a Neanderthal may seem cute—it is certainly fanciful but, as we have seen, among humans' fellow-apes there are species still around today that raise similar problems.

The insight of Prometheus—that humans, wretched as we are, have a divine spark about us and don't deserve to die—is now, in the light of present knowledge, extensible to apes. Perhaps an appropriate response would be to stretch our thinking to new limits—embracing apes in the same moral community as ourselves. Peter Singer has led a campaign along those lines. Even before he became Professor of Bioethics at the Center for Human Values at Princeton University, he was one of the most widely read contemporary philosophers, thanks to his championship of animal rights and his disarmingly candid arguments in favour of euthanasia, abortion, and even infanticide. He was one of the first to argue that the boundaries of the genus *Homo* should be extended to include chimpanzees. This is a perfectly reasonable proposal. Genetically, the differences between us and chimps are small enough to justify it, by comparison with the way taxonomists draw the limits of other genera. Moreover, as we have seen, the overlap in capacities and behaviour between humans and chimps seems to grow as research makes progress. And in the world of hominid fossils, the case for redefining the genus is in any case extremely strong: the current chaos is evidently irrational and needs sorting out. Singer goes on to advocate the admission of some apes to some rights: not, of course, to all, since the right to vote, for instance, would be unlikely to interest them, but rights of the sort that humans of similar interests and capacities might exercise. These would include rights to live in peace, without expropriation from their domain or subjection to capture, torture, or experimentation of kinds which inflict pain or deprivation. People who recognize how many common features of our lives and psyches unite us to non-human apes respond warmly to Singer's advocacy. There are, however, two reasons for hesitating.

Firstly, it is impossible to devise a coherent argument in favour of incorporating non-humans in our moral community, without excluding some humans. If one extends some rights to apes, one must obviously do so not because of the species in which they happen to be classified, but on the grounds that they evince appropriate characteristics. If the same rule were applied in establishing human entitlement to rights, many people who do not possess the same characteristics would be excluded. Singer himself is candid about that. A creature who can feel pain ought to be protected from it; in the case of a human so inert or comatose or paralysed as to be capable of feeling nothing, the right to be exempt from pain is strictly meaningless. Why should a vegetative human have human rights, while a passionate, self-conscious, reflective creature like a chimp or bonobo can be imprisoned, used to try out dangerous drugs, or sliced up for science? The concept of human rights crumbles. Singer says he can see no reason for rights applied solely on the basis of classification in the human species.

Secondly, human rights, if they are genuinely human, must be indivisible—applicable to all humans. If they are selectively applied, they are not rights but in a sense privileges. But if we apply them to all humans, we do an injustice to excluded apes, who qualify—by fair tests—for some of them; and if we extend them to other species they cease to be human rights. The process of extending rights, as Singer eagerly acknowledges, would be open-ended. Once apes were in the moral community on the grounds that they resemble humans, we should have to confront the case for extending them again, this time to monkeys, on the grounds that monkeys resemble apes, and so on until we reached Bertrand Russell's oysters. And what then? We should be in a pickle similar to that of E. M. Forster's missionary, convinced by Brahminical interlocutors that grace and redemption could not decently be limited to humans, but that monkeys, too, should 'have their collateral share of bliss'. And, the Brahmins then said, 'What about oranges, crystals and mud?'

We could acknowledge the overlaps between apes and humans, without making the mistake of treating them as identical, by extending some human rights to apes; but elasticity means two-way stretch: more exclusions as well as more inclusions. While apes are admitted to our moral community, on the grounds that they have quasi-human characteristics, humans who lack the qualifying characteristics of consciousness and sentience will lose the most basic right of all—the right to life: those who are unborn, or moribund, or deeply cognitively disabled. Characteristics-ism will replace species-ism. Non-human apes would be exempt from vivisection, but unborn human babies could be cannibalized for spare parts. Such a conclusion is logically tenable—but is it morally sustainable, humanly tolerable?

We could perhaps simply stop worrying about our inability to define human nature, and adopt cosmic altruism —putting other species first. It might be ecologically salutary, but it would not be practical: no society could be that unselfish and survive. Or we could attempt to re-think our concept of humankind from scratch—tearing up the long record of achievement, writing off all the blood and effort that has gone into making us a little more humane than our predecessors. We may have to embark on such re-thinking anyway. If current trends in philosophy and science do not force us into it, future developments might. What if we succeed in producing beings genetically or mechanically endowed with the moral equivalence of humanity, programmed for consciousnes? If apes and Neanderthals don't make us revise our concept of humankind, maybe cyborgs will.

CHAPTER 5

POST-HUMAN FUTURES?

Humankind in the Age of Genetics and Robotics

In her latest book, the leading neuroscientist Susan Greenfield has re-fashioned the debate. As we still do not understand what consciousness is, she argues, there is little point in speculating about the prospects of implanting it in robots. The foreseeable outlook—which can fairly be called a danger—is of another kind. Being human will be different in a body constantly modified by genetic upgrades, a family in which sex and reproduction are no longer linked, a 'reality' which is 'virtual', relationships that are kaleidoscopically mutable, and an electronically engineered environment constantly re-crafted by your mood. According to Greenfield, technology will erode individuality, as it replaces memory and makes experience vicarious. We will lose hold of our continuous narratives of our lives, our sense of what makes us different from each other. The effect will be de-humanizing, because, according to Greenfield, 'personalization' is what turns the brain into a mind—and personalization is the effect of accumulating experiences and interactions. We will revert to a reptile-like phase of evolution, in 'this state of sensory oblivion, stripped of all cognitive content and bereft of self-consciousness'. It will resemble states we now enjoy on a temporary basis, such as bovine contentment or mindless torpor—the pleasures of the slob, the spaced-out, and the couch potato. But these pleasures can be savoured only when they are rare. 'By incessantly stimulating neuronal connections into constrained configurations,' Greenfield writes, 'the new technologies might be jeopardising the very existence of Human Nature, permanently.'

Individuality eradicated by computer? Humankind replaced by a race of nerdish sociopaths, living vicariously wired lives? The frightening thing about Susan Greenfield's future is that we will not need a Frankenstein or a Dr Moreau, genetic engineering or artificial intelligence, to inaugurate a post-human era. It is implicit already in our appetite for self-immersion in ersatz reality. We seem doomed to self-transformation into something unrecognizable as human by present standards. How can we head this future off—and the other post-human futures which technology dangles before our fears? We have to confront the promise and menace of genetics and robotics.

The Genetic Crucible

In lectures in Dublin in 1944, Erwin Schrödinger speculated about what a gene might look like. He predicted that it would resemble a chain of basic units which connected like the elements of a code. The nature of DNA as a kind of acid was not yet known. Schrödinger expected a sort of protein, but the idea he outlined galvanized the search for 'the basic building-blocks' of life. When James Watson saw X-ray pictures of DNA, he realized—by his own account—that it would be possible to discover the structure Schrödinger had predicted. He joined Francis Crick's project in Cambridge and got a great deal of help from a partner-laboratory in London, where work by Rosalind Franklin, who contributed vital criticisms of Crick's and Watson's unfolding ideas, helped build up the picture of what DNA was really like. The Cambridge team, in effect, appropriated Franklin's work without her permission and without letting her know what they were doing; doubts over whether they had played fairly could not, however, detract from the importance of the outcome. It took a long time for that importance to emerge fully, with the growing realization that genes in individual genetic codes were responsible for some diseases. The double-helix shape, in which the molecular structure of DNA twists, has proved mesmerizing: it has become one of the most powerful, widely reproduced, and instantly recognized icons of our time. The possibilities opened up by its discovery have had an inebriating effect, evident among enthusiasts tempted into the speculation that our genes could account for many or even all kinds of behaviour, which could be regulated by changing the code. Similarly, the codes of other species can be modified—and are being modified—to obtain results which suit humans: producing bigger plant foods, for instance, or animals designed to be more beneficial, more palatable, or more packagable for human food.

Progress in research in these fields has been so rapid that it has raised the spectre of a world re-crafted, as if by Frankenstein or Dr Moreau, with unforseeable consequences. People now have the power to make their biggest intervention in evolution yet: selecting 'unnaturally' not according to what is best adapted to its environment, but according to what best matches agendas of human devising. 'Designer babies' are already being produced in cases where genetically transmitted diseases can be prevented. The prospect that some societies will want to engineer human beings along the lines prescribed in former times by eugenics is entirely likely. Morally dubious visionaries are already talking about a world from which disease and deviancy alike will have been excised.

We shall return to these fears in a moment. Meanwhile, genetic research is doing us a lot of good: alerting people with a genetic predisposition to certain diseases, enabling them to seek treatment, take precautions, or exercise proper prudence. Further research will unlock new therapies: modifying the genes of sufferers from genetically transmitted diseases. Most of this will carry promises without threats—without exhuming Frankenstein or building post-human monsters. One of the first benefits of the code-cracking which has elicited the human genome is a revolution—a new, dazzling clarity—in our concept of what it means to be human. We now have, for the first time ever, a scientifically viable—because quantifiable—definition of humankind. A human is a creature with a genome, measurably defined. If your DNA differs from that of other creatures defined as human in these terms by less than 0.1 per cent, then you have achieved the ambition we ascribe to the ape: you are human too.

The trouble is that genetic research, which provides this definition, also shows how feeble and arbitrary the distinction between humans and other animals is. We share nearly 95 per cent of our genome with chimpanzees—not much less than we share with each other. The difference which makes us human is a barely significant difference.

One route of escape from the dilemma is to abandon the search for a biological definition altogether and to revert to the quest of Chapter 3: attempting to define human-ness culturally. Human is as human does. Membership of the community is what counts. This could prove to be a highly pernicious doctrine. Communities are fluid. The historical trajectory of the concept of humankind, as this book shows, has been one of more-or-less progressive self-enlargement. Would it ever have been right to attempt to halt the process and say that membership of the human community is now closed? Would it be right to do so now? After all, it might be argued, our understanding of what it means to be human is now so generous, so sweeping, that further elasticity is not needed. Yet there are still some categories unassimilated by the process so far, or in danger of exclusion—either from classification as fully human or from the enjoyment of corresponding 'human rights', or both. As we have seen, some groups of hominids and extinct humanoid species have yet to be satisfactorily classified. The fate of these long-vanished creatures may be regarded as a purely academic matter; but there are unresolved questions about the status of an important group of human beings who are still with us: the unborn. And similar questions may extend to the terminally ill, the victims of deep coma or severe mental impairment, the old and demented, and others near death.

The doctrine that human-ness is socially definable—that our human nature 'kicks in' only when we relate to other humans—is used and, one suspects, may really have been devised, or adopted by some of its advocates, as a justification for abortion or the immolation of human embryos for the sake of scientific experiment. By any reasonable standards, unborn babies must count as human: what other species could they possibly belong to? Humans in embryo are a form of human life: there is, in the present state of our knowledge, no known way of differentiating a moment when the creature in the womb can be said not yet to be alive from another, subsequent moment at which that creature is alive.

23. Superficially savage, in a manuscript compiled under missionary guidance in the early colonial period, Aztec training methods for oblates actually resemble the ascetic practices admired in the Church at the time and are surely calculated to invite approval. Flesh is mortified by beating and piercing with cactus spikes; the trained priest is tonsured.

24. Bernard's romanticised engravings, with their antique conventions for depicting bodies and drapes, illustrated Lahontan's account of Huron courtship rituals: women supposedly chose sexual partners, signifying assent by blowing on a suitor's torch. The allure of free love excited philosophers attracted to noble savagery.

25. William Hodges, artist of Cook's second voyage, launched the tradition of painting Omai as an aristocratic hero. Cook regarded the islander as 'a downright blackguard' and the ship's surgeon thought he wanted to leave Tahiti, where he was a captive, because people laughed at his wide nostrils. But in London society he acquired a reputation for 'good breeding' which reinforced romantic notions of noble savagery.

26. 'Peter the wild boy' was found, naked and silent, in the woods of Hanover in 1725 as a young teenager. Under learned tuition in London, he learned to mime polite greetings. Defoe defended him from the charge of mere idiocy as an example of how 'mere nature' needs society to civilize it. He spent much of his life tethered to prevent escape, popularly reported in the Gentleman's Magazine as 'more of the Ouran Outang species than of the human.'

27. The anthropologist Margaret Mead, shown here with her native collaborator, Fa'amona, claimed to record supposedly liberated sexual practices among Samoan adolescents in the 1920s. Fortified by Freudian critiques of repression, she was one of the last great propagators of the romantic myth of savagery as superior to civilisation. (left)

28. In the 1860s, the Harvard Anthropologist Louis Agassiz travelled in Brazil in the hope of vindicating his theory that 'mixed races' were physically inferior and incapable of producing fertile offspring; if true, this would have strengthened the case for classifying humans into a number of species. He insisted his female subjects strip for photography, claiming that the shapes of breasts varied according to race. The photographs seem scientifically inert, yet evince unmistakable prurience. (below)

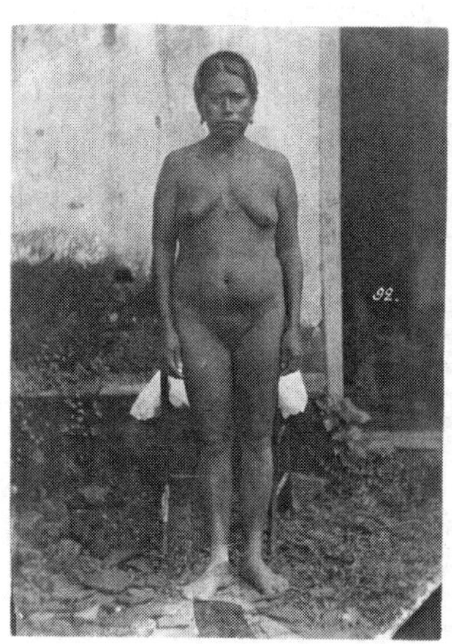

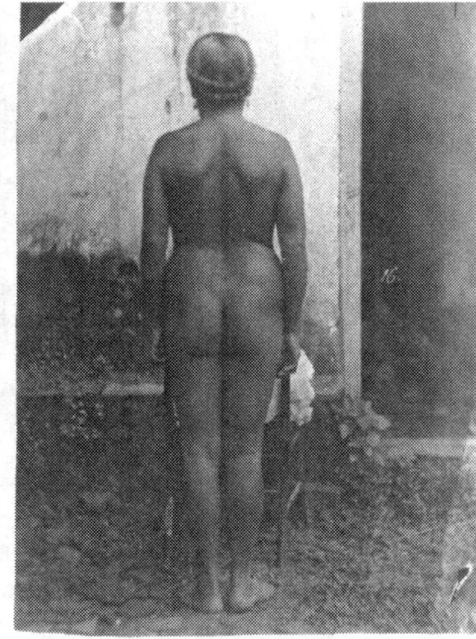

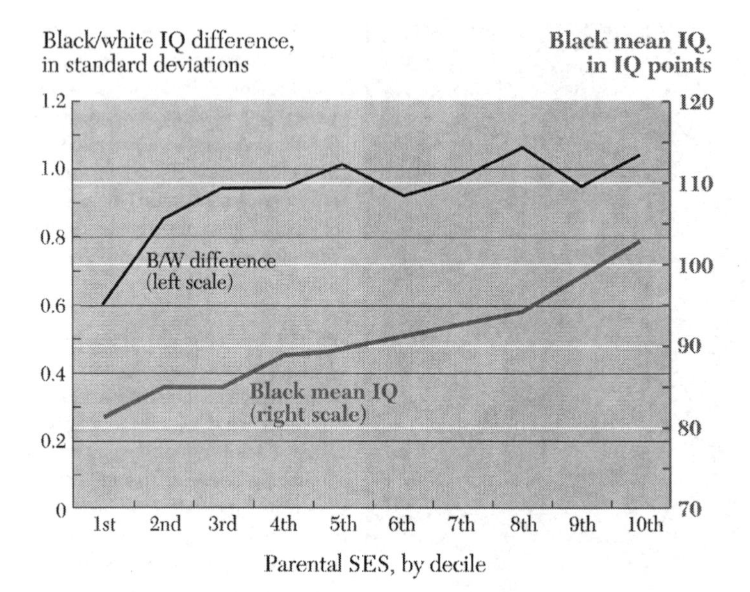

29. A diagram from The Bell Curve: Intelligence and Class Structure in American Life (1994), by Richard Herrnstein and Charles Murray, purports to show that, although the intelligence of offspring increases with parents' economic and social status, the difference between blacks and whites remains the same.

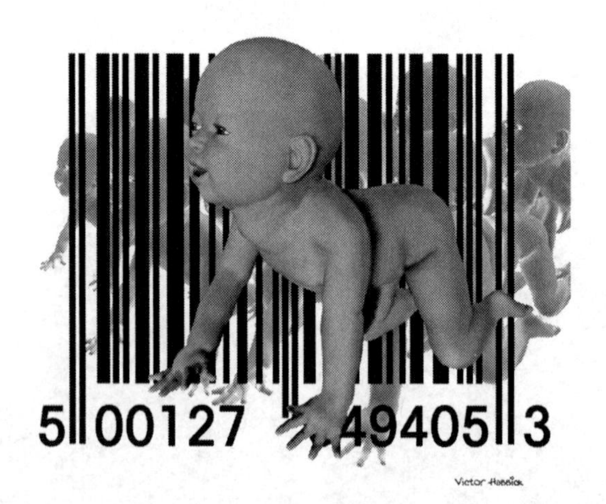

30. Cloning as commodification: bar-coded babies, in this computer-generated conceptual-art image, represent the distaste the idea of cloning evokes. Genetic engineering replaces inherited with injected characteristics. Cloning is a further stage: 'genetic manufacture' of babies grown, perhaps, from patented cells.

It is at best weasely, at worst wicked, to contrive a definition which excludes the unborn or which relegates them to an inferior category; and it is hardly more rational to do so than to contrive—as other justifiers of iniquities have done in the past—a definition which excludes or relegates Jews, say, or blacks. The unborn are the great, remaining underprivileged minority of our day, whom we can exterminate without a qualm. It is impossible to justify the hecatomb of foetuses—the massacre of innocents practised daily in today's 'developed' world—without discarding the notion of humankind we have built up, so painfully and triumphantly, in the course of our history, and re-defining our community of fellow creatures in some other way: as, for instance, the community of those possessing some postulated qualification, usually called, in the context of the abortion debate, 'personhood'.

I should stress that my focus on the problem of abortion is intended only as an illustration of the difficulty of sustaining our present concept of humankind. I am not seeking to influence debate on the laws governing abortion or to align myself with any lobby. Unhappily, I know that the law is not there to serve justice or reason or logic or morals. Considerations of convenience, custom, culture, and, in democratic societies, of consensus are also part of the law-making process. Although I would be glad to see stricter abortion laws, I do not advocate them: my delight would flow from the fact that such laws would be evidence of enhanced sensibilities and deeper wisdom in our society, not because they would make life harder for women struggling to cope with unwanted pregnancies. Nor would I presume to criticize people who have abortions or partners who condone them, or medical and nursing staff who take part in the operations. In all cases I am aware of, practitioners of abortion sincerely believe that they are making moral choices, and that unborn babies—at least, at early stages of embryonic development—are not fully human, in the sense of being fully entitled to human rights, because they lack personhood, or 'are part of their mothers' bodies' or are excluded from rights the rest of us have because of some other alleged deficiency. The resemblances between these arguments and those used to justify victimization of other human groups in the past are chilling; and the sincerity with which they are advanced is no guarantee of their goodness. But they do mean that it would be unreasonable and improper to bandy around some of the accusations common in the anti-abortion lobby, such as that practitioners are evil, or 'murderers', or ought to be punished rather than pitied. Anti-abortion legislation, in present circumstances, would criminalize people of good conscience, without saving all unborn babies' lives. More effective, in the long term, would be a debate informed by science, reason, and research into embryonic development. This might advance, if not settle, debate on whether there is a moment beyond conception which can properly be identified as the start of human life.

The Selfish Genie: Building Superhumans

Shakespeare's horse-painter 'would surpass the life in limning out a well-proportioned steed'. It was a time when every artist was an idealist and merely to reflect nature was not enough: artists had to exceed nature to excel in art. For scientists, it is an even greater temptation. Experiments to 'improve' on naturally occurring species—to modify them to suit human purposes—have at least ten thousand years of history behind them. Humankind's first great successful challenge to evolution occurred when farmers and herders began to improve on natural selection by culling and cross-breeding—a kind of 'unnatural selection' designed to produce animals tamer or tastier than those produced by evolution, crops more prolific and easier to grow. The results included some of the most spectacular transformations science has ever effected. From grasses indigestible to human stomachs, for instance, came the great staple foods of the world: wheat, rice, maize, barley, and other grains. Whole zones of the Earth, which were once forest and scrub-land and swamp, became stamped with the geometry of fields and cities.

The most advanced hybridization techniques endued their practitioners with seemingly magical powers of 'improvement' on nature. The greatest plant-life magus of all time was surely Luther Burbank, whose hybridization experiments in the 1880s made him the Frankenstein of the vegetable world. He delighted in bizarre, eye-catching inventions: a white blackberry, a stoneless plum, a new fruit that was half-plum, half-apricot, a giant potato, and a thousand new species all told. He was notorious for the profligacy of his methods, which involved the immolation of thousands of unsuccessful hybrids. It is not surprising that this model—audacity to breed, voracity to destroy—should have been tried out as a technique for improving humans. The dream of re-engineered humans, who would improve God's handiwork, was an ancient one. Plato's proposals for a perfect society rested, in part, on the assumption that it should consist of perfect individuals: the best citizens should be encouraged to reproduce, the children of the dim and deformed exterminated to stop them breeding. These repellent proposals were the outcome of a common observation probably already thousands of years old in Plato's day: the importance of heredity, which was widely assumed long before genetic theory produced a convincing explanation of why—for example—some looks, skills, quiddities, diseases, and deficiencies run in families.

Plato's idea was shelved for many centuries. In nineteenth-century Europe and North America, however, it revived under the influence of racism—which blamed heritable deficiencies of character for the supposed inferiority of non-whites—and a form of Darwinism, which suggested that the supposed advantages of natural selection might be helped along by human agency. In 1885, Darwin's cousin, Sir Francis Galton, proposed what he called eugenics: the human species could be perfected by the elimination of undesirable mental and moral qualities, and this could be achieved by selectively controlled fertility. 'If a twentieth part of the cost and pains were spent on measures for the improvement of the human race that is spent on the improvement of the breed of horses and cattle, what a galaxy of genius might we not create!' wrote Galton in 1865. It was, at first, perhaps, a programme for universal improvement, endorsed by Darwin himself—despite differences with his cousin over the technicalities of heritable characteristics—a few years later. But racism came to dominate the debate and inform the eugenicists' programme. 'Eugenics co-operates with . . . Nature,' was Galton's conclusion in 1904, 'by securing that humanity shall be represented by the fittest races.'

Within a couple of decades, this became one of the orthodoxies of the age. In early Soviet Russia and parts of the USA in the same period, the right of marriage was denied to people officially classified as feeble-minded, criminal, and even (in some cases) alcoholic. By 1926, compulsory sterilization of people in some of these categories had been adopted in nearly half the states of the USA. The eugenic idea was most enthusiastically adopted in Nazi Germany, where its logic climaxed: the best way to stop people breeding was to kill them. Anyone in a category the state deemed genetically inferior, including Jews, gypsies, and homosexuals, was liable to extermination. Meanwhile, Hitler tried to perfect what he thought would be a 'master race' by selective breeding between people supposedly of the 'purest' German physical type. The state housed big, strong, blue-eyed, blonde-haired human guinea-pigs for experimental copulation. Children born of their unions did not, on average, seem any better or any worse qualified for citizenship, leadership, or strenuous walks of life than other people.

Nazi excesses made eugenics unpopular for generations. But the concept is now back in a new guise: 'genetically engineered' individuals can be reproduced. Indeed, 'sperm banks' of semen donated by people allegedly of special prowess or talent have long been available to potential mothers willing to shop for a genetically superior source of insemination. The isolation of particular genes associable with various inherited characteristics has, moreover, made it theoretically possible for characteristics thought undesirable to be 'filtered out' of the genetic material that goes into a baby at conception. At present, the consequences are incalculable.

They could include the 'post-human future' independently predicted by contrasting intellectuals. Brian Appleyard is an admirably, uniquely polymathic journalist who has waged a personal crusade against scientism. Francis Fukuyama is an academic 'primarily interested in culture and economics', who is one of the world's most influential spokesmen for democratic, liberal capitalism. In contrasting books—the former's jittery, colourful, anguished, the latter's calm, judicious, practical—both recognize the difficulty of 'staying human in the genetic future'. 'Biotechnology,' Fukuyama warns, 'will cause us in some way to lose our humanity. . . . Worse yet, we might make this change without recognizing that we had lost something of great value. We might thus emerge on the other side of a great divide between human and posthuman history and not even see that the watershed had been breached.' I suspect this is true, for a reason demonstrated by this book: we do not know what humankind means; we do not know what it is that makes us human; so naturally, we will not be aware of losing it.

Much of the opposition to genetic research is ignorant, bigoted, sensationalist, or based on sectional assumptions and interests. In 1995 a 'coalition' of self-styled 'religious leaders' brought together by Southern Baptists in the USA signed a declaration of opposition to the patenting of genes on the grounds that they were the property of the real creator: God. This was laudable as well as laughable: patenting of genes is a serious evil of our times. It gives big business crushingly excessive power in the marketplace for food and medicine, whereas we desperately need to encourage the more environmentally friendly initiatives typical of small farmers, artisanal producers, and individual or small-scale medical practitioners and carers. But the God-centred argument sounds theologically illiterate—everything in creation is God's, but that does not stop some of us from 'owning' bits of it—as well as narrowly fanatical. We need secular arguments to convince non-religious people of the need to control the effects of the genetic revolution.

The arguments are hard to separate from the emotional reactions the subject arouses. One of the most powerful icons generated by biological engineering is the widely reproduced photograph of a mouse with an expression reminiscent of a human ear growing out of its back. There is already a rhesus monkey with jellyfish genes in all its body's cells. The monsters falsely anticipated in legend, and unnecessarily accommodated in medieval sages' taxonomies of nature, can now be created: chop-and-change creatures, assembled like Lego-men from lab-grown spare parts. In the future imagined by Francis Fukuyama, however, the post-humans will not be physical monsters so much as cocktails of well-shaken personality characteristics, aptitudes, and dispositions. Wielding a metaphor surely indebted to the ear-backed mouse, he writes, 'The ultimate question raised by biotechnology is, What will happen to political rights' when, in effect, we can 'breed some people with saddles on their backs and others with boots and spurs.' Fukuyama points out that we already use drugs to tune the personalities of people we find disruptive or uncongenial or hard to control, or to enhance the powers—stamina, sexual prowess, wakefulness, resistance to stress—of people who can afford the cost. It is practically inconceivable that genetic engineering will be left unexploited for similar purposes.

It will also become part of the armoury with which we strive for immortality. Our obsessive desire to prolong our lives seems odd when so many of them are empty or filled only with meretricious comforts or rewards. But the temptation to curtail or regulate ageing processes is strong. It is beyond belief that science would forgo any opportunity of selling a protracted existence to paying clients. Biotechnology will help us realize the nightmare of Marcel Ainé. In *La carte,* his surrealist short story, written during the war, the rich buy up the ration cards that allot people time.

Like immortality, infertility is one of the obsessions of our time: both are grounds for experiments in human cloning. Infertility is one of the many afflictions re-evaluated in societies becoming unused to frustration. It is a normal condition, which its victims have faced in the past but which now seem intolerable to sufferers. The right to have a family is obviously not meant to mean that everyone who wants children must have them: only that governments must not forbid them, as, in some countries, they have done by compulsory sterilization programmes or legislation proscribing procreation. In any case, human cloning seems an unnecessarily risky and elaborate way of countering infertility: existing procedures are now so effective that the cloning technology is supererogatory. The WHO, UNESCO, and the European Parliament have made explicit declarations condemning human cloning as unethical. Many countries have banned it. The most promising or threatening initiatives are in the private sector—and not always from reputable sources. A few days after Christmas 2002, a blaze of publicity revealed unsubstantiated claims that human children had been successively cloned by a private firm. The financial backing came from a sect which claims that the human race was founded by extraterrestrial visitors to Earth. The technology is sure to be developed—various clinics and companies are committed to perfecting it—but the availability of alternative treatments, the ubiquity of moral disquiet, and the taint of association with weird lobbies on the irrational fringe all make it unlikely to be widely used in confronting infertility.

Therapeutic cloning is different. This means breeding embryos to ransack them for useful cells. The ethics of this—farming human lives for spare body parts—are repellent, but most governments, so far, have shown little unease about it. We know that mice can be redesigned by adding and deleting genes to the embryonic cells. In cases involving Combined Immune Deficiency Syndrome, human embryos have been treated by the same method. Genetic screening, even without any further progress in positive interventions to modify the genetic makeup of human beings, will be a big challenge to the course of evolution. A woman can pick and choose the most perfect specimens—or those she most fancies—from among as many embryos as she cares to produce. The rest can be destroyed. This is already common practice for cystic fibrosis, which is associated with a single gene and for a range of other conditions, including multiple sclerosis, a form of intestinal cancer, retinitis pigmentosa (which causes blindness), and Tay-Sachs disease (which attacks the central nervous system). In a famous case in 2000 a woman in St Paul had a baby called Adam whose genetic disposition to anaemia was corrected by embryo selection, so that replacement cells from his body could be given to his sister. As I write these lines in May 2003, the UK courts have cleared a proposal for a similar intervention in England. Single-gene diseases are almost all very rare, but as the technology develops, all heritable diseases could be dealt with by a similar method. A further stage, the development of which seems inevitable, is screening for diseases which have a genetic component: in effect, this will destroy embryos merely because they might develop diseases, which could normally be perfectly well controlled by conventional therapies, including remedies as simple as dieting. This really is a cure which kills. At present, we do not know whether the result will be a healthier species: recoded DNA may have side-effects; cloned specimens of non-human creatures have had chequered health histories.

Last time governments thought they had a chance to 'improve' people by breeding for heritable traits, the results included state-sponsored eugenics programmes in which 'imbeciles' were sterilized and ultimately, in some countries, 'undesirable' types were exterminated. This time around, however, the 'new eugenics'—according to defenders of a genetic free market—will be entirely benign. Single-gene disorders, or those caused by a limited number of identifiable genes—will vanish soon. Freedom will increase, because people will no longer be condemned by genetic predispositions to particular diseases, they will no longer have to spend their lives fighting them off: people who would otherwise be killed by cholesterol will be able to eat what they like; those vulnerable to cancer will be able to smoke. The anxiety and depression which afflict modern society will be alleviated as people are liberated from worrying about their health. The genes responsible for more serious mental disorders could be spotted and gutted. Anxieties, however, about the implications for humankind's future focus not on manipulation for health so much as for character and abilities: the 'menace' of 'designer babies'.

Personality genes are, in principle, a credible proposition. There is a very rare gene which increases appetite. There is another which effectively protects people from alcoholism. The search for an 'intelligence gene' is probably bootless, but a 'super-mouse' called Doogie was engineered by gene-enhancement in Princeton and MIT labs in 1999: he was significantly better at memory and adaptation tests than other mice. Other genes do seem to affect moodiness, propensity to anxiety, and disposition towards serious mental disorders, including manic depression and schizophrenia. The 'criminal gene' is not a fantasy. In 1991 a criminal conviction—a murder committed in the course of a robbery—was appealed on the grounds that the defendant came from a family with an antisocial history and could therefore be supposed to be the victim of a heritable and perhaps curable condition. In 1993 a study was conducted at Nijmegen into a local family of notoriously violent criminal propensities. All the males were mentally retarded; all exhibited violence when angry; between them, their crimes covered threatening behaviour, rape, stabbing, attempted murder, arson. They also shared a genetic abnormality which led to high levels of adrenaline release. But members of the family who inherited the same gene were able to behave in normal, socially productive ways. One was happily married and employed in a stable job. Some people really do inherit a genetic mutation which causes abnormally high levels of adrenalin release. Everyone knows the effect: adrenaline is nature's lubricant for combat.

According to one point of view, the prospects opened by these facts are no more worrying than those of genetically improved future health. The political environment of the new eugenics—we are assured—will be liberal. Genetically perfected humankind will be different from eugenically perfected humankind because there will be no losers—or, at least, few losers, apart from the unborn immolated for experiment and the probably very few victims of attempted cures or modifications which may go wrong. The evils of the old eugenics arose from the involvement of states, with their uniformist agendas. The power of individual choice will perpetuate diversity and make genocidal abuse of the technology impossible. This seems excessively sanguine. One of the dangers of the GM revolution is precisely the threat it poses to freedom. Ultimately, genetic modification will have to be a state monopoly. Individual competition would negate its effects. Alternatively, if it were left to market forces, the rich could buy up the best genes, creating Fukuyama's 'genetic overclass': the equalizing effects of the lottery of life would be obliterated. We should be condemned to life in a caste system. If, as Francis Crick claims, 'Almost all aspects of life are engineered at the molecular level', morality gets crowded out of mind. The more genes there are to predispose behaviour, the less responsible the individual is for achievements, crimes, follies, virtues, vices.

According to a popular assumption, the power of genes has settled one of recent history's most significant scientific controversies: significant, that is, in its direct impact on ordinary lives. This is the 'nature versus nurture' debate: the conflict between those who believe that character and capability are largely inherited and therefore unadjustable by 'social engineering' and those who believe that these qualities are induced by experience and that social change can therefore affect our moral qualities and collective achievements for the better. Broadly speaking it is a left/right conflict, with supporters of social radicalism ranged against those reluctant to make things worse by ill-considered attempts at improvement. The controversy crystallized in the late 1960s in the rival reports of Arthur Jensen at Berkeley who claimed that 80 per cent of IQ is inherited (and, incidentally, that blacks are genetically inferior to whites) and Christopher Jencks and others at Harvard who also used IQ statistics to argue the opposite case. The same argument, supported by the same sort of data, was still raging in the 1990s, when Richard J. Herrnstein and Charles Murray published *The Bell Curve*, arguing that society has a hereditary 'cognitive elite' and 'underclass' (in which blacks are disproportionately represented). They predicted a future of cognitive class-conflict.

Although the IQ 'evidence' was highly unconvincing, anxieties of this kind were fed by genetic research, which seemed to confirm that more of our makeup is inherited than we have traditionally supposed. Meanwhile, sociobiology, a 'new synthesis' devised by the ingenious Harvard entmologist Edward O. Wilson rapidly created a scientific constituency for the view that differences between societies are determined by evolutionary necessities and that societies can be ranked accordingly. Two fundamental convictions have survived in most people's minds: that individuals make themselves, and that society is worth improving. Nevertheless, the power of genes threatens to eliminate our confidence in our freedom to equalize the differences between individuals and societies: this encourages the prevailing conservatism of the early twenty-first century. Even psychiatry—which for most of the twentieth century was a discipline shaped by the conviction that experience shaped personality—has lost confidence in our ability to reach the mind, except through drugs. Now the fashion is for 'evolutionary psychology', which emphasizes the role of genes. During the twentieth century, the claim that much human motivation is subconscious challenged—and successfully challenged—traditional notions about responsibility, identity, personality, conscience, and mentality. Now the challenge is being mounted by genetics.

Genetics also invalidates existentialism—the characteristic way in which people of my generation, educated in the 1950s and 1960s, thought about what it means to be human, accepting our existence between birth and death as the only immutable thing about us and tackling life as a project of self-realization, of 'becoming': who we are changes as the project unfolds. 'Man,' Jean-Paul Sartre said, 'is only a situation' or 'nothing else but what he makes of himself . . . the being who hurls himself toward a future and who is conscious of imagining himself as being in the future.' For Sartre, self-modelling was not just a matter of individual choice: every individual action is 'an exemplary act', a statement about humankind, about the sort of species each of us wants us to be. Yet there is no objective way of informing such a statement. God does not exist, everything is permissible, and 'as a result man is forlorn, without anything to cling to. . . . If existence really does precede essence, there is no explaining things away by reference to a fixed . . . human nature. In other words, there is no determinism, man is free, man is freedom.' This very nearly became the orthodoxy of a generation in rebellion against every other orthodoxy. Now, as a means of coping with the predicament of being human, it looks valueless.

Meanwhile, a deeper paradox is at work: the genetic revolution is already altering the way we think about what it means to be human, by nudging us as we drift into a materialist conception of ourselves. According to Francis Crick, the co-discoverer of the significance of the double helix, 'the soul has vanished, along with most metaphysics'. It is true that the soul has vanished from many people's account of themselves. In part this is the result of prejudices nourished by science, which tends to privilege materialism because only matter is measurable and detectable to the senses. A trend in philosophy has filleted mind out of our lexicon: we are left with mindless brain. Neurological research makes it ever easier to see thought as a process in which synapses fire and proteins are released, matter grinding against matter. Doubts about 'mind–body dualism' were best answered by an aphorist in *Punch* in 1855: 'What is mind? No matter. What is matter? Never mind.' Now the question arises, are we genes or genie? Is there a djinn in our bottle?

The conviction that the self is in some sense spiritual is remarkably resilient—considering how strange it seems by the canons of common sense. There is no test for spirit; but the sense of it is too strong and widespread to be dismissed. We do not know what we should do in a future without it, because we have neither record nor memory of such a past. The disenchantment of life could lead to disenchantment with life. The imaginative discovery that life is animated by spirit was probably one of our ancestors' first thoughts. People who reject it call it 'primitive' but it is really profound; only a supple mind could manage it. Materialism was surely the first philosophy, the genuine primitivism. To abandon spirit would be to revert to the presumable mental world of the australopithecines or even earlier creatures. The post-human future may be technologically sophisticated, but in one respect it is likely to resemble the pre-human past.

Whether 'mind' is just a fancy word for 'brain' or whether the soul—as Carlyle quipped—'is a kind of stomach' are questions about one kind of material reality: living matter; but, according to an equally searching challenge, minds may not even be necessarily organic: they could be merely mechanical. Artificial intelligence research has reinforced the suspicion. If a machine could successfully replicate human thought, the mechanical nature of intelligence would have been demonstrated. Or the machine might have to be classed as human, inspirited in the same way as ourselves. Like progress in genetic research, the speculation has threatened to revive fear of the Frankenstein myth. Intelligent machines would surely outsmart humans. They could replicate without our help. They could become the next step in evolution. All our thoughts could be downloaded, as if between computers, and the immortality cloning may fail to confer could be achieved by information technology. In fact, human intelligence is probably fundamentally unmechanical: there is a ghost in the human 'machine', whereas computers only 'quasi-think' within formal, closed systems. Except to those prejudicially committed to a mechanistic definition of thought, there is no reliable test of whether a machine is thinking like a living creature. (Alan Turing famously devised one such test: can you tell from its conversation that it is a machine? But this is fallacious. A machine could apply conversational rules without understanding what was said.) On the other hand, computers can certainly extend minds and change them in ways we have only begun to explore. They have revolutionized the world, in combination with microtechnology, which made small computers viable (the early specimens made in Manchester and Harvard were the size of small ballrooms) and telecommunications technology, which made it possible to link computers along telephone lines and by radio signals, so that they could exchange data. By the early twenty-first century, it really had become possible to re-imagine the world as a 'global village', where every part of the globe could be in virtually instant contact with every other part. This has been both a disaster and an opportunity: 'information overkill' has glutted minds and, perhaps, dulled a generation; but the Internet has multiplied useful work, diffused knowledge, and served freedom. A new psychology is being forged in cyberspace. Though intelligent engines may be unconvincing sci-fi heroes, the impact of the infotec revolution on how we think and therefore, *a fortiori*, how we think about ourselves has barely begun. Its future trajectory is incalculable.

There are two ways of escape from the bleakness of supposing that we have to forfeit our delusions of freedom and accept that we are what we once thought other animals were: automata or soulless brutes. The first is to realize that the dilemma is false: we may be products of evolution and prisoners of our genes; but that does not make us unfree. We still re-shape ourselves through experiences we endow with meaning. The second route lies through our self-re-classification as part of nature: but a nature which is full of enchantment, and in which no life is merely brutish.

The most menacing thing about genetics is not that it will change humankind but that it will change life, of which humankind is just a tiny, ephemeral part. We do not know what it means to be alive any more than we know what it means to be human—but we do have a sense of what it is like from our own experience of it: it means to suffer and to sense suffering and to feel sorrow and misery and guilt and pain—and to know goodness and love and joy and peace only because we also experience the loss of them and the absence of them and the opposite of them. No wonder we question human nature: whether thanks to Eve or evolution, that is part of the deal. And whatever human nature is, dilemmas and dissatisfaction are among the results of it.

The Need to Re-think

We have to face challenges to our concept of humankind and, where they are valid, confront the consequences for some of our dearest concepts: human rights, human dignity, human life—ideas for which, in the recent past, we've begun to re-shape states, re-fashion laws, and fight wars. We might attempt a new characterization along the lines proposed by Justin Stagl. He has suggested that humankind should be defined dynamically in terms of a combination of characteristics in shifting combinations, capable of changing over time without losing touch with tradition: these include, most importantly, a 'biological heritage' of ill-developed instincts and reflexes which make humans open to change; our legacy of transformation 'from a biologically determined to a socio-culturally determined being' for whom culture has become natural and which can, in a sense, be measured in our diversity of cultures; and, above all, our 'utopian potential', our vocation to transcend our failures and defects, 'to strive to attain superhuman goals and avoid the inhuman'. Our human nature may be mired in the animal continuum, but our sense of 'better nature', at least, is unique.

There is still no agreement about what 'human nature' is—what, beyond trivial or temporary features of our physiologies or our cultures that happen to have been thrown up by history or evolution, is common to and exclusive to the creatures we recognize as human. Human nature, if it is proper to speak of such a thing, is not fixed: it has changed in the past and could change again. Its continuity with the natures of other animals is part of its fluidity. We are like other life—only, in some respects, more so: more so, in respect of capacity to develop culture and change the way we live; more so, in respect of some skills and some weaknesses, some of our capacity for good, some of our capacity for evil. How much our nature has to change before our descendants cease to be human is a question we are not yet ready to answer. In this respect it resembles the question about when, in the course of evolution, our ancestors became human—which is also unanswerable at the present stage of our thinking and knowledge.

That humans are uniquely rational, intellectual, spiritual, self-aware, creative, conscientious, moral, or godlike seems to be a myth—an article of faith to which we cling in defiance of the evidence. But we need myths to make our irresoluble dilemmas bearable. And our claims for our nature are more: not mere myths but also aspirations, still waiting to become true. By the standards of the Utopian hopes Justin Stagl identifies, we are bestial creatures indeed; but those glimpses of self-elevation to a genuinely different category—to the level of angel, or god, or comic-book super-hero—give us precious self-dissatisfaction, which we can work at and build on. If we were uncompromising mythbusters, we would tear up our human rights and start again: re-think what we mean by human life and human dignity. For now, if we want to go on believing we are human, and justify the special status we accord ourselves—if, indeed, we want to stay human through the changes we face—we had better not discard the myth, but start trying to live up to it.

Further Reading

INTRODUCTION AND CHAPTER ONE

On species theory, I follow the line well expressed in D. Hull, 'A Matter of Individuality', *Philosophy of Science,* xlv (1978), 335-60 (see also his *Science as a Process* [1988]) and M. Ereshefsky, *The Poverty of the Linnaean Hierarchy: a Philosophical Study of Biological Taxonomy* (2001). For opposing views, see for example R. Wilson (ed.), *Species: New Interdisciplinary Studies* (1999).

K. P. Oakley, *Man the Tool-Maker* (1975) is the classic statement for differentiating humans on the basis of tool-use; W. Kohler, *The Mentality of Apes* (latest edn, 1999) shows that it is mistaken. J. Goodall, *The Chimpanzees of Gombe* (1986) represented a decisive breakthrough in this, as in many other respects concerning the supposed differences between apes and humans.

The best introductions to the primatological approach to the study of human nature are D. Morris, *The Naked Ape* (1967), A. Desmond, *The Ape's Reflexion* (1979), J. Diamond, *The Third Chimpanzee* (1992), F. de Waal, *The Ape and the Sushi Master* (2002), and F. de Waal (ed.), *Tree of Origin* (2001), which contains several important papers. J. Goodall, *Open Window* (1990) is an autobiographical account of her work. D. Fossey, *Gorillas in the Mist* (1984) is an enthralling account of her work in Rwanda. On primate culture see also W. McGrew, *Chimpanzee Material Culture* (1992) and I. DeVore (ed.), *Primate Behaviour* (New York, 1965). C. B. Stanford, *The Hunting Ape: Meat Eating and the Origins of Human Behaviour* (1999) argues from a primatological perspective for meat-eating as the essential ingredient of human nature; though this is not a position I should care to endorse, the book is a highly commendable updating of a theory formerly thought discredited.

The Chomsky Reader (1998) is a useful guide to his work. My quotations come from N. Chomsky, *Knowledge of Language* (1986). On Kanzi, see K. R. Gibson and T. Ingold, *Tools, Language and Cognition in Human Evolution* (1993). On Lucy see M. K. Temerlin, *Lucy, Growing up Human: A Chimpanzee Daughter in a Psychotherapist's Family* (1976). For primatological insights on the origins of language see R. Dunbar, *Grooming, Gossip and the Evolution of Language* (1997) and C. T. Snowdon, 'From Primate Communication To Human Language', in deWaal (ed.), *Tree of Origin*, pp.193-227. S. Pinker, *The Language Instinct* (1994) is an entertaining, powerfully reasoned work of popular linguistics.

R. Hinde (ed.), *Non-verbal Communication* (1972) and D. Morris, *Manwatching* (1977) are particularly important on non-verbal communication. J. T. Bonner, *The Evolution of Culture in Animals* (1980), D. Griffin (ed.), *Animal Mind–Human Mind* (1982), and S. Walker, *Animal Thought* (1983) are pioneering works on the evolutionary context of gesture and language. See also B. King, *The Information Continuum: Evolution of Social Information Transfer in Monkeys, Apes and Hominids* (1995).

For an engaging essay on avian communications, see A. Powers, *Birdtalk* (2003). On parrots see particularly I. M. Pepperberg, *The*

Alex Studies: Cognitive and Communicative Abilities of Grey Parrots (2000). On the 'people without art' see W. Grainge White, *The Sea Gypsies of Malaya* (1909). Some paintings by non-human apes are reproduced in D. Morris, *The Biology of Art: A Study of the Picture-Making Behaviour of the Great Apes and its Relationship to Human Art* (1962). On fire see R. W. Byrne, 'Social and Technical Forms of Primate Intelligence,' in deWaal (ed.), *Tree of Origin*, pp. 145-72 (the orang-utan story is on p.167) and W. C. McGrew, 'Chimpanzee Material Culture: What Are its Limits and Why?' in R. Foley (ed.), *The Origins of Animal Behaviour* (1991), pp. 13-22. On fire generally see J. Goudsblom, *Fire and Civilisation* (1994).

On the context of Descartes' views see T. Hastings, *Man And Beast in French Thought* (1936) and L. D. Cohen, *From Beast-Machine to Man-Machine* (1941). For English ideas see K. Thomas, *Man and the Natural World* (1983). For Singer's opinions see P. Singer and S. Reich, *Animal Liberation* (1990). For context on animal-rights thinking, see T. Benton, *Natural Relations* (1993), R. G. Frey, *Interests and Rights: The Case Against Animals* (1980); M. Midgeley, *Beast and Man* (1980). For the quotation from Montaigne, see his *Essays* (1893), ii, 145.

My quotation from M. Tomasello is from *The Cultural Origins of Human Cognition* (1999), p. 22.

T. D. Price and J. A. Brown (eds.), *Prehistoric Hunter-Gatherers* (1985) and J. D. Lewis-Williams, *Discovering South African Rock Art* (1990) are indispensable for undertanding the shamanism of cave-painters. M. Eliade, *Le Chamanisme* (1951) is the now widely repudiated anthropological classic which defined debate. For Kuhn's work see H. Kuhn, *On the Track of Prehistoric Man* (1955) and *The Rock Pictures of Europe* (1967). See J. Dow, *Shamanism* (1990) for a modern survey. On totemism, C. Lévi-Strauss, *Totemism* (1962) is an ingenious and convincing critique. For Chinese material, I rely on D. Boddie, *Chinese Thought, Society, and Science* (1991) (see especially pp. 310-27) and J. Needham et al., *Science and Civilisation in China* (1956—in progress) (my quotations are from vol. ii, 17, 56, 374, 575). J. -C. Schmitt, *Le saint lévrier* (Paris, 1979) tells the St Guinefort story and that of the grasshoppers of Segovia is in W. Christian, *Local Religion in Sixteenth-century Spain* (1981), pp. 30-1.

CHAPTER TWO

J. Gray, *Straw Dogs: Thoughts on Humans and Other Animals* (2002) is a profound and disturbing reflection on the impossibility of defining the animal frontier. It can usefully be set alongside other master-works of scientifically informed pessimism about human nature: A. Koestler, *The Ghost in the Machine* (1976) and R. Fox, *The Search for Society: Quest for a Biological Science and Morality* (1989). The rumination of Adriaan Kortland is quoted by Adrian Desmond, *Ape's Reflection*, p. 229.

H. W. Janson, *Apes and Ape Lore in the Middle Ages and Renaissance* (1952) is the insuperable guide to the history of ideas about apes. For Jobson the best text is S. P. Gamble and P. E. H. Hair (eds.), *The Discovery of River Gambra by Richard Jobson* (1999). A facsimile edition of E. Tyson, *Orang-outang, sive Homo Silvestris, or, the Anatomy of a Pygmie [1699]*, A. Montagu (ed.), appeared in 1966. M. Hodgen, *Early Anthropology in the Sixteenth and Seventeenth Centuries* (1964) is an admirable survey. The best edition in translation of the Popol Vuh is D. Tedlock (ed.), *Popol Vuh: The Mayan Book of the Dawn of Life* (1985). On Baldaeus, I have relied on A. de Jong, *Peter Baldaeus, afgoderyn der Oost Indische Heydenen* (1917) (the story told is on p. 24). The cited press reports on pygmies occurred in The *Guardian*, 9 Jan. 2003 and *The Week*, 31 May 2003. On pygmies see C. Turnbull, *The Forest People* (1961). For the quotation from Augustine see *City of God*, book XVI, section 8.

On medieval slavery, C. Verlinden, *L'Esclavage dans l'Europe médiévale*, 2 vols. (Bruges, 1955) is the standard work, supplemented by J. Heers, *Esclaves et domestiques au Moyen-âge dans le monde méditerranéen* (1981) and W. D. Phillips, *Slavery from Roman Times to the Early Transatlantic Trade* (1985). For Portugal I have relied on A. C. de C. M. Saunders, *A Social History of Black Slaves and Freedmen in Portugal, 1450-1550* (1982). R. Segal, *Islam's Black Slaves* (2002) is a good introduction, which supplied my quotations from Suhaym and ibn Rabah (pp. 47, 49).

On the origins of racism see H. E. Augstein (ed.), *Race: The Origins of an Idea, 1760-1850* (1996) and M. Banton, *Racial Theories* (1987). N. L. Stepan, *The Idea of Race in Science* (1980), C. Bolt, *Victorian Attitudes to Race* (1971), L. K. Uper (ed.), *Race, Science and Society* (1975), and G. W. Stocking (ed.), *Race, Culture and Evolution* (1968) are outstanding studies. For useful selections from sources, see P. J. Kitson (ed.), *Theories of Race* (1999). Lawrence published his *Lectures on Physiology, Zoology and the Natural History of Mankind* in 1819. My quotations are from the 1823 edition. The quotation from Herbert comes from Thomas, *Man and the Natural World*, p. 42. On representations by painters see B. Smith, *European Vision and the South Pacific* (1960). The comparative material on China comes from S. -T. Leong, *Migration and Ethnicity in Chinese History: Hakkas, Pengmin and their neighbours* (1997).

The work cited by D. Pick is *Faces of Degeneration* (2001), from which the quotation from Maudslay is taken. On Lamarck, H. G. Cannon, *Lamarck and Modern Genetics* (1960) can be recommended. In this chapter, quotations from Darwin are all from *The Descent of Man* (1871). The approach of S. Morton is exemplified in *Crania Americana* (1839), and that of J. C. Nott and G. Gliddon in *Indigenous Races of the Earth, or New Chapters of Ethnological Enquiry* (1857). Crawfurd and Hunt are dealt with superbly in T. Ellingson, *The Myth of the Noble Savage* (2001). On Gobineau, M. D. Biddiss, *Father of Racist Ideology: Social and Political Thought of Count Gobineau* (1970) is a useful introduction. On the implications of evolutionary biology for racism the work of E. Mayr is fundamental: see *What Evolution Is* (2001). Mayr's contribution to species theory is also helpful in locating humankind in relation to other creatures. See *Animal Species and Evolution* (1963) and *Towards a New Philosophy of Biology* (1988).

CHAPTER THREE

Adam Kuper, *The Chosen Primate* (1994) is a candid attempt to face the problems of defining humankind culturally. R. Bartlett, *Gerald of Wales* (1982) is the best introduction to that writer whose works on Ireland are edited in J. O'Meara (ed.), *The History and Topography of Ireland* (1982). My pages on medieval ethnography are based on those on the same subject in F. Fernández-Armesto, *Millennium* (1999). On wild men, see R. Berheimer, *Wild Men in the Middle Ages: A Study in Demonology* (1970) and T. Husband (ed.), *The Wild Man: Medieval Myth and Symbolism* (1980).

The paragraphs on Columbus are based on F. Fernández-Armesto, *Columbus on Himself* (1991). The images of Aztec life described in the text are from Codex Mendoza, of which there are good editions by J. Cooper Clark (1932) and F. Berdan and P. Anawalt (1992).

On the tradition generated by New World encounters, the works of A. Pagden are fundamental, especially *The Fall of Natural Man* (1987) and *European Encounters with the New World from Renaissance to Enlightenment* (1994). For the context of Montaigne, see F. Lestringant, *Cannibals* (2000). H. Fairchild, *The Noble Savage* (1928) concentrates on English texts but remains unbettered on the subject. M. Newton, *Savage Boys and Wild Girls: A History of Feral Children* (2002) summarizes and excels all previous work on the subject. On bonobo sexual life see F. de Waal, 'Apes From Venus,' in *Tree of Origin*, pp. 39-68.

CHAPTER FOUR

Quotations from Darwin on the Fuegians are from the appendix to his *Voyage of the Beagle* (innumerable editions since its first appearance in 1840). On the cave of Atapuerca see J. L. Arsuaga, *El collar del Neanderthal* (1999). On Lucy, D. C. Johanson and M. A. Edey, *The Beginnings of Humankind* (1981) and *Lucy's Child* (1990) are insiders' accounts of the discovery and its importance. Lively accounts of the use of DNA in tracing human ancestry are L. L. Cavalli-Sforza, *Geni, popoli e lingue* (1996) and B. Sykes, *The Seven Daughters of Eve* (2001). M. H. Brown, *The Search for Eve* (1990) is a dispassionate narrative. On Neanderthals, C. Stringer and C. Gamble, *In Search of the Neanderthals* (1993) is an outstanding introduction and E. Trinkaus and P. Shipman, *The Neanderthals* (1993) a fascinating history of modern thought on the subject. P. Mellars, *The Neanderthal Legacy*, 1996, and J. Shreeve, The *Neanderthal Enigma*, 1995, survey the evidence. For the case 'against' the 'humanity' of Neanderthals, see P. Lieberman, *Uniquely Human: The Evolution of Speech, Thought and Selfless Behaviour* (1991) and E. Trinkhaus, 'Bodies, Brawn, Brains and Noses: Human Ancestors and Human Predation', in M. H. Nitecki and D. V. Nitecki (eds.), *The Evolution of Human Hunting* (1987), pp. 107-45, with the rebuttal in Stanford, *The Hunting Apes*, p. 133. W. Golding, *The Inheritors* (1955) is a fiction inspired by the speculation that humans might deliberately have exterminated the Neanderthals. J. Reader, *Missing Links: The Hunt for Earliest Man* (1988) is a good history of palaeoanthropology. On Mozu see F. de Waal, *Good Natured* (1996), pp. 6-7.

There is a recent edition of F. Galton, *Essays in Eugenics* (1985). M. S. Quine, *Population Politics in Twentieth-century Europe* (1996) sets out the context. M. B. Adams (ed.), *The Well-born Science* (1990) is a collection of important essays. D. J. Kevles, *In the Name of Eugenics: Genetics and the Uses of Human Heredity* (1985) is a useful history of eugenics. M. Kohn, *The Race Gallery* (1995) studies the rise of racial science. N. Stepan, *Race and Science in the Nineteeth Century* (1982) and G. W. Stocking, *Victorian Anthropology* (1987) are outstanding studies; two chapters in N. L. Stepan, *Picturing Tropical Nature* (2001) bring the former up to date in key respects and outline the importance of environmentalist thinking on racial ideas. C. Clay and M. Leapman, *Master Race* (1995) is a chilling account of the Nazi project, as is B. Muller-Hill, *Murderous Science: Elimination by Scientific Selection of Jews, Gypsies and Others, Germany 1933-45* (1981).

P. Singer, *Writings on an Ethical Life* (2001) and *Unsanctifying Human Life: Essays on Ethics*, H. Kuhse (ed.) (2002) are introductory selections. For his championship of apes see P. Cavalieri and P. Singer (eds.), *The Great Ape Project: Equality Beyond Humanity* (1993).

CHAPTER FIVE

The latest thinking is in S. Greenfield, *Dreams and Shadows: How Twenty-first Century Technologies are about to Transform Our Thoughts, Feelings and Personalities* (2003). I am grateful to the author for showing me a chapter in progress. My allusions are to B. Appleyard, *Brave New Worlds: Genetics and the Human Experience* (1999) and F. Fukuyama, *Our Post-human Future: Consequences of the Biotechnology Revolution* (2002). For a rational defence, see D. Galton, *In Our Own Image* (2001). For more alarmed reflections see M. Rees, *Our Final Century* (2003).

My quotations from Francis Crick are from *What Mad Pursuit* (1990) and *The Astonishing Hypothesis* (1995). On the attempt to quantify intelligence, S. J. Gould, *The Mismeasure of Man* (1982) is a brilliant, scholarly polemic. R. Lewontin et al., *Not in Our Genes: Biology, Ideology and Human Nature* (1984) is an important collection in resistance against biological determinism, while E. O. Wilson, *On Human Nature* (1978) is the best statement on the other side of the question. D. R. Hofstadter, *Godel, Escher, Bach: An Eternal Golden Braid* (1981) is the most brilliant—though ultimately unconvincing—apologia for 'artificial intelligence' ever penned. K. Hafner and M. Lyon, *Where Wizards Stay Up Late* (1996) is a lively history of the origins of the Internet. On philosophical dimensions, see R. Trigg, *Ideas of Human Nature* (1988) and A. O'Hear, *Beyond Evolution: Human Nature and the Limits of Evolutionary Explanation* (1997). For Stagl, see J. Stagl, 'Anthropological Universality: On the Validity of Generalisations about Human Nature' in N. Roughley (ed.), *Being Humans: Anthropological Universality and Particularity in Transdisciplinary Perspectives* (2000), pp. 25-46.

Index

Index

Index

Index

Index

Index

Index

Index

Lightning Source UK Ltd.
Milton Keynes UK
UKOW03f2332261113

221871UK00003B/6/P